D1165448

Lecture Notes in Physics

The Lecture Notes in Physics

The series Lecture Notes in Physics (LNP), founded in 1969, reports new developments in physics research and teaching – quickly and informally, but with a high quality and the explicit aim to summarize and communicate current knowledge in an accessible way. Books published in this series are conceived as bridging material between advanced graduate textbooks and the forefront of research to serve the following purposes:

• to be a compact and modern up-to-date source of reference on a well-defined topic;

• to serve as an accessible introduction to the field to postgraduate students and nonspecialist researchers from related areas;

• to be a source of advanced teaching material for specialized seminars, courses and schools.

Both monographs and multi-author volumes will be considered for publication. Edited volumes should, however, consist of a very limited number of contributions only. Proceedings will not be considered for LNP.

Volumes published in LNP are disseminated both in print and in electronic formats, the electronic archive is available at springerlink.com. The series content is indexed, abstracted and referenced by many abstracting and information services, bibliographic networks, subscription agencies, library networks, and consortia.

Proposals should be sent to a member of the Editorial Board, or directly to the managing editor at Springer:

Dr. Christian Caron
Springer Heidelberg
Physics Editorial Department I
Tiergartenstrasse 17
69121 Heidelberg/Germany
christian.caron@springer.com

Hans-Jürgen Borchers   Rathindra Nath Sen

Mathematical Implications of Einstein-Weyl Causality

2006

 Springer

Authors

Hans-Jürgen Borchers
Georg-August University, Göttingen
Faculty of Physics
Institute of Theoretical Physics
Friedrich-Hund-Platz 1
37077 Göttingen
E-mail: borchers@theorie.physik.uni-
goettingen.de

Rathindra Nath Sen
Ben-Gurion University of the Negev
Faculty of Natural Sciences
Department of Mathematics
Beer Sheva 84105
Israel
E-mail: rsen@cs.bgu.ac.il

H.-J. Borchers and R.N. Sen, *Mathematical Implications of Einstein-Weyl Causality*,
Lect. Notes Phys. 709 (Springer, Berlin Heidelberg 2006), DOI 10.1007/b11809043

Library of Congress Control Number: 2006930923

ISSN 0075-8450
ISBN-10 3-540-37680-1 Springer Berlin Heidelberg New York
ISBN-13 978-3-540-37680-4 Springer Berlin Heidelberg New York

Springer is a part of Springer Science+Business Media
springer.com
© Springer-Verlag Berlin Heidelberg 2006

Typesetting: by the authors and techbooks using a Springer LATEX macro package
Cover design: WMXDesign GmbH, Heidelberg

Printed on acid-free paper   SPIN: 11809043   54/techbooks   5 4 3 2 1 0

Dedicated to Waltraut Borchers and Alice Sen

Preface

While working on our common problems, we often used our meetings to discuss also the fundamentals of physics and the implications of new results in mathematics for theoretical physics. Out of these discussions grew the work presented in this monograph. It is an attempt at answering the following question: What mathematical structures does Einstein–Weyl causality impose on a point-set M that has no other structure defined on it? In order to address this question, we have, first of all, to define what precisely we mean by Einstein-Weyl causality – that is, to provide an axiomatization of this notion. It may be remarked that we were led to the axiomatization given in Chap. 3 by physical intuition, and not by mathematical analogy.

Next, we show that our axiomatization defines a topology on the point-set M, and that the topological space M is uniformizable. We then show that, if the topology of M is first-countable and its uniformity is totally bounded, then the order completion of the uniformity[1] has a local differentiable structure. Examples show that the conditions of first countability and total boundedness are sufficient, but not necessary. However, the methods we have developed so far are not applicable to more general situations. Roughly speaking, we can summarize our results as follows: *Subject to the above caveat, spaces satisfying Einstein–Weyl causality are densely embedded in spaces that have locally* but not necessarily globally *the structure of differentiable manifolds.* Physical intuition can give no more, since a finite number of measurements cannot distinguish between infinitely-differentiable and (continuous but) nowhere-differentiable structures.

Motivations for axiomatizing a physical theory may vary. Carathéodory "axiomatized" classical thermodynamics in order to understand temperature and entropy in terms of integrating factors of Pfaffian forms. Wightman's axiomatization of relativistic quantum field theory was a response to the situation that a mathematically inconsistent theory gave results in perfect

---

[1] Order completion is slightly different from uniform completion; however, the difference may be disregarded at this stage.

agreement with experiment. Closer to our own endeavours, the Soviet school, following A. D. Alexandrov, has axiomatized relativity theory (see [44, 45]) in the spirit of Hilbert's sixth problem, and Schutz has developed "Minkowski space-time from a set of independent axioms stated in terms of a single relation of intermediacy or betweenness" [95].

The fundamental difference between these axiomatizations inspired by the theory of relativity and our work appears to be the following. In the works cited, a definite physical-mathematical structure is taken as "given"; in one case it is relativity theory, in the other Minkowski space. The aim in each case is to axiomatize a given mathematical structure in its entirety. By contrast, we seek to isolate a fragment (which we call Einstein-Weyl causality) from the totality of structures called relativity theory, attempt a precise definition of this fragment only, and investigate what further mathematical structures it forces on the underlying point-set. The difference is strikingly illustrated by the fact that Einstein-Weyl causality can be defined on discontinuua such as $\mathbb{Q}^2$. The relationship with the differentiable structure of space-time is more subtle; naively speaking, we are trying to unearth *mathematical* consequences of a *physical* principle. The question we investigate is not inspired by Hilbert's sixth problem, but rather by Alexandrov's observation [1] that the interiors of double cones provide a base for the usual topology of Minkowski space, by Zeeman's work that "Causality implies the Lorentz group" [132], by Cantor's assertion that "the very essence of mathematics is its freedom" (see [38], pp. 3–4), and by Wigner's consequent query that if mathematics is the free creation of the human mind, then how does one explain "The unreasonable effectiveness of mathematics in the natural sciences" [128]?

The mathematical level required of the reader is that of the graduate student pursuing a problem in mathematical physics. For the physicist who is interested in applications, perhaps the most significant result of the present work is that the notion of causality can effectively be extended to discontinuua.

The few references to physics may be disregarded by mathematicians without loss of continuity. Sections 8.2.1, 11.2 and the unsolved problem discussed in Chap. 9 may be of particular interest to them. The extension of our results to infinite-dimensional spaces remains a major problem.

Every chapter begins with a paragraph or a section that motivates the chapter and gives an indication of the results sought and established. Appendix A gives a fairly detailed summary of the basic theory of uniformities and uniformization. Appendix B gives the definitions and results on fibre bundles and G-structures that are needed in the text. Appendix C brings together the axioms and the special assumptions. A List of Symbols is provided. The Index is mainly an index of terms; it is not comprehensive.

Göttingen   *Hans-Jürgen Borchers*
Pardes Hanna   *Rathindra Nath Sen*
May 2006

Contents

1

Introduction

In every textbook on the theory of relativity it is assumed, without any discussion, that space-time is a differentiable manifold, possibly with singularities. Between a point-set and a differentiable manifold there is an enormous gap, and we felt that physics itself could contribute to narrowing this gap. In experimental physics, one can make only a finite number of well-separated measurements, and therefore it is natural to start with a discrete set of points as a candidate for space-time. As one cannot put an upper bound on the number of measurements, this set cannot be finite; it must, at least, be countable. Next, as one cannot place a quantitative limit on experimental accuracy, one has to admit the "density" property that between any two points on a scale lies a third. Finally, one has to ask how one arrives at the continuum. In short, we felt that it should be possible to start from point-sets and find conditions (axioms) – motivated by physics – which would allow us to construct a topological manifold from the point-set. If this turned out to be true, one could become more ambitious and look for conditions which would imply the differentiability of the manifold.

The problem with such a project is that while the goal is clear enough, the starting point is rather less so. Since we had something like the background for relativity theory in mind, the goal had to be an ordered manifold where the order is causal and is defined by local cones.[1] It would have been too much to expect that one could begin with the definition of these local cones, since a cone is, in some sense, a complex higher-dimensional object which should be constructed from simpler objects. The simplest ordered sets are those that are totally ordered. From the experience of Minkowski space, one knows that a causal order determines two classes of totally ordered subsets, namely the paths of light rays and those of freely-falling particles. These subsets have the density property that between any two distinct members of one lies a third. A good example of such a set is provided by $\mathbb{Q}$, the set of rational numbers.

---

[1] *Local cones* are the intersections of cones with the neighbourhoods of their vertices.

H.-J. Borchers and R.N. Sen: *Mathematical Implications of Einstein–Weyl Causality*,
Lect. Notes Phys. **709**, 1–6 (2006)
DOI 10.1007/3-540-37681-X_1   © Springer-Verlag Berlin Heidelberg 2006

One class of such objects is defined in Chap. 3; the objects are called *light rays*, short for space-time paths of light rays, which is what they would be in Minkowski space. It is possible to define light cones in terms of light rays, but one has to introduce additional conditions to ensure that the cones so defined be proper cones. The global behaviour of such cones can be rather strange. Only the properties of local cones can be specified in advance. The local cones should be convex and their boundaries should consist only of light rays passing through the vertex (no higher dimensional faces, such as one finds in pyramids, are allowed). That the cones should be "well-behaved" only locally reflects the idea that gravitational lensing, which is a global effect, should be permitted.

Forward and backward cones lead trivially to the definition of double cones (order intervals).[2] The collection of double cones that are are "small enough" and have the desired "nice properties" can be used to define a topology. The definition of the nice properties and the construction of the topology is given in Chap. 4. It turns out that the nice properties defined for the small double cones are nice enough to ensure that the topology is not only Hausdorff, but also allows the definition of sufficiently many continuous real-valued functions. The order intervals with the nice properties that will be used in this text will be called *D*-intervals.

From double cones one can construct quadrilaterals bounded by four light rays. Moreover, if the double cones are *D*-intervals, then by using only light rays one can construct natural, order-preserving homeomorphisms between the opposite sides of light ray quadrilaterals. Using these homeomorphisms one can derive certain homogeneity properties of the entire space, which go a long way towards showing that the space looks everywhere the same. These results are established in Chap. 5. However, to do so one has to carry small *D*-intervals along light rays, which is only possible if light rays have overlapping coverings by the basic sets of the topology. This is an additional requirement that has to be imposed.

Nothing so far excludes the possibility that the space we are considering is totally disconnected (like $\mathbb{Q}^2$); we have to ask whether or not our space can be completed. Fortunately the topology defined in Chap. 4 is nice enough to make this possible. In order that a completion be possible it is necessary for the space to carry either a metric or a uniform structure. A metric structure is too restrictive for our purposes, but our spaces do carry the relevant uniform structure. This concept is explained, and the result extablished, in Chap. 6. Having completed the space, one has to extend the order to the completed space. This is also done in Chap. 6.

Chapter 7 is in the nature of an aside, and discusses some properties of spaces in which light rays are complete. An example shows that there are infinite-dimensional spaces in which light rays are complete, but the

---

[2]   A double cone is the intersection of a forward and a backward (light) cone.

space itself is not. However, no finite-dimensional space with this property is known.

The threads are pulled together in Chap. 8, in which several major topics are covered. The first is the construction of timelike curves, which allows one to introduce coordinates in D-intervals. With these one completes the proof that the space does indeed look everywhere the same, and answers key homotopy questions. The final result in this direction shows that the interior of a D-interval in a finite dimensional space is homeomorphic with $\mathbb{R}^n$ for some n, and therefore has a differentiable structure. This is followed by a discussion of the possible global differential structure; here the assumption of isotropy plays an essential role. The fact that such an additional assumption is necessary for the differentiable structure to be global suggests that the exotic differentiable structures on $\mathbb{R}^4$ may be of limited physical interest.[3] The last topic of this chapter is the construction of covering spaces if the original one is not simply connected. The main point of this extension is to show that the order structure can be extended to the covering space. The converse, namely that the restriction of an ordered covering space is also ordered, need not be true.

The timelike curves we constructed in Chap. 8 were coordinate dependent. Try as we might, we could not construct natural ones, owing to a peculiar problem that arises in ordered spaces. This is the cushion problem, which is discussed in Chap. 9. Chapter 10 discusses two works that may be said to "bracket" ours, and calls attention to some that are less closely related. Chapter 11, Concluding Remarks, discusses some physical, mathematical and philosophical questions arising from or related to our work.

Our enterprise ends by introducing a conformal structure. A suggestion for going further and defining the Weyl projective structure is given at the end of Sect. 11.1. This could be the starting-point for defining the Riemannian structure, following the work [32] of Ehlers, Pirani and Schild.

The above is a brief outline of the programme of this monograph, which describes several layers of mathematical structure that follow from a fundamental physical principle. Most of these mathematical structures and their interrelations were first analysed by Weyl [116], and a brief account of these structures is given in Chap. 2. The notion of causality is discussed in Sect. 1.1, and the key role of Alexandrov's observation is explained in Sect. 1.2. This Introduction ends with some remarks on space-time at the Planck scale.

---

[3] For more information on exotic $\mathbb{R}^4$'s, we refer the reader to the very readable article by Freedman [40]. A more technical account may be found in Gompf [42]. The books by Lawson [69], Bourguignon and Lawson [13] and Atiyah [3] cover the application of Yang-Mills theory to geometry and topology in greater detail. Donaldson's original article is [25]. We shall not discuss these questions in this monograph.

1.1 Causality as a Physical Principle

The term *causality* has been used in 20th-century physics in two senses; a general sense indistinguishable from determinism in pre-relativity physics, and a more specific one that began to take shape in the nineteen-twenties (and will be explained below). An example of the first of these may be found in Weyl's book ([115], p. 207):

> "...we have ten "principal equations" before us, in which the derivatives of the ten phase-quantities with respect to the time are expressed in relation to themselves and their spatial derivatives; that is, we have physical laws in the form that is demanded by the **principle of causality**" [emphasis in the original].

Weyl apparently meant to distinguish between causal or deterministic and non-deterministic time evolution. If one assumes that there is no effect without a cause,[4] a non-deterministic evolution can only mean an evolution subject to external influences that defy precise accounting.

In the 19th century the relation between cause and effect acquired a new twist owing to the emergence of Cauchy's (initial-value) problem for hyperbolic differential equations. This problem arose quite naturally in Lord Rayleigh's *Theory of Sound* [105] and Maxwell's electrodynamics [76]. It turned out that the solution of the initial-value problem for a later time depended only upon a finite part of the initial surface [47]. The special theory of relativity gave a quantitative turn to the notion that causes preceded effects: the speed of light set a limiting velocity for the transmission of a signal. Using this fact (together with the "conservation of probability"), Kramers [65] and de Kronig [67] showed (in 1926–27) that, in the scattering of light by atoms, the forward scattering amplitude at any frequency is related to an integral, over all frequencies, of the total absorption cross-section (see, for example, [8]). This relation is now known as the Kramers-Kronig relation. It was the starting-point of the theory of dispersion relations, and introduced the term *causality conditions* into the lexicon of physics. Since the 1940's, a further terminological refinement seems to have taken place. The fact that no physical signal can propagate faster than the speed of light has been called *macrocausality*. Its major consequence in quantum theory – that observations at spacelike separations cannot influence each other – is called *microcausality*. In relativistic quantum field theory[5] the preferred term is *local commutativity*; Fermi fields anticommute rather than commute at spacelike separations. All of these terms describe what are, essentially, physical principles.

---

[4] This was a metaphysical belief of the enlightenment. As we are concerned only with scientific questions and not with philosophical ones, we do not have to contest this assumption. A nice discussion of the relation between inferring causes from effects and vice versa is given in Jordan [56].

[5] The central role of causality in quantum field theory is fully brought out in the book by R. Haag [46].

As suggested earlier, the terms causality and determinism are no longer synonymous in physics, with the former having clear mathematical overtones. In current usage, a causal system often (but not always) means one of the hyperbolic type, the paradigm for which is the wave equation (which describes, inter alia, the propagation of sound and electromagnetic waves). A causal system is certainly deterministic; however, there exist deterministic systems, such as those governed by the equation of heat conduction (or diffusion), that are not of the hyperbolic type, and therefore not "causal" in the sense described above.

Our use of the terms *causal* and *causality* will conform to the restricted sense of the last paragraph.

1.2 Causal Structures as Primary Objects

In physics, it is taken for granted that space-time is a differentiable manifold, except for speculative works on discrete space-times (the 1967 work of Kronheimer and Penrose [66], which we shall discuss later, is a notable exception). As stated earlier, the question we wanted to address was, roughly speaking, the following: Is it possible to trace the topological and differential structures of space-time to something that could be interpreted as a *physical* principle? The hint came from the following:

It had been observed by A. D. Alexandrov [1] in 1959 that the interiors of double cones in Minkowski space provide a base for its (standard) topology. Double cones can also be defined in Newtonian space-time. In Newtonian physics, there is no upper limit to the speed of propagation of a signal; that is, one may assume the velocity of light to be infinite. A "light cone" in Newtonian physics is therefore a 3-dimensional hyperplane $t = \text{const} = a$ which separates the future part of space-time, $t > a$, from the past part $t < a$. A double cone is a slice $a \le t \le b$. The interiors of these slices also form the basis for a topology on space-time, $\mathbb{R}^4$. However, this topology is not the standard one; it is not even Hausdorff (see Fig. 1.1).

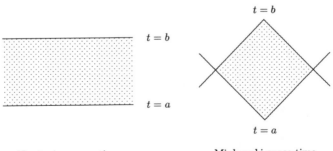

Newtonian space-time   Minkowski space-time

Fig. 1.1. Double cones in two-dimensional space-time

In this connection one should also mention the work of Zeeman [132], who showed, without any assumption of continuity, that the group of automorphisms of Minkowski space that preserve the causal order is the inhomogeneous Lorentz group plus the dilations.[6]

1.3 Space-time at the Planck Scale

We end this Introduction with a remark concerning space-time at the Planck scale (ca. 10^{-33} cm).

Most authors believe that the quantization of general relativity will present new difficulties at the Planck scale. Due to Heisenberg's uncertainty relations, distances smaller than the Planck length will probably make no physical sense. It has been suggested that discretization would be a way out. Our approach shows that discretization is consistent with causality, but it is doubtful whether it offers a solution to the problem. Unless one is able to fix a fundamental length, discretization will have to be random or fuzzy, ideas that have not found much favour. Some authors have suggested the alternative of a quantized space-time, i.e., one in which the coordinates do not commute. The classical, i.e., commuting coordinates should then appear as a limit, rather like classical physics being a limit of quantum mechanics. What happens with causality at this scale is not known. As usual physics is likely to break down in black holes, quantization of general relativity may be very different from the standard quantization of fields.

---

[6] The "fine topology" that he later developed for Minkowski space in [133] does not appear to have been investigated further.

2

Geometrical Structures on Space-Time

The chief aim of this chapter is to define the geometrical structures, global and local, that are associated with the notion of causality. The work described later in this volume is the exploration of an abstraction from this notion. We shall begin with a review of the relevant global geometrical structures on $\mathbb{R}^n$, and then proceed to the local structures. We shall follow Hermann Weyl's approach [115], modified as necessary to incorporate later developments, in the discussion of geometrical structures. A few terms used by Weyl and still in use have changed meanings, and we shall point these out. Most of this material will be familiar to geometers and relativity theorists, but perhaps less so to others.

In the following, we shall also explain why we have chosen the title *Mathematical Implications of Einstein-Weyl Causality* for this volume; that explanation will simultaneously provide the setting for our endeavour. As the term "Einstein–Weyl Causality" is meant also as a tribute to Weyl,[1] we shall give a few quotations to convey an idea of Weyl's thinking on the subject.

2.1 Global Structures on $\mathbb{R}^n$

We begin with the most basic definitions.

By $\mathbb{R}^n$ we shall understand the Cartesian product of n copies of $\mathbb{R}$, usually with the product topology. The following geometrical structures are defined *globally* on $\mathbb{R}^n$:

1. **The Affine Structure:** On p. 12 of *Space–Time–Matter* [115], Hermann Weyl wrote: "My real object... will be to single out **translations** among possible congruent transformations. Starting from the conception of translation I shall then develop Euclidean geometry along strictly axiomatic

---

[1] In defining the conformal structure (see Sect. 2.2.3), Weyl provided a framework for the study of causality independently of the notion of length.

H.-J. Borchers and R.N. Sen: *Mathematical Implications of Einstein–Weyl Causality*,
Lect. Notes Phys. **709**, 7–14 (2006)
DOI 10.1007/3-540-37681-X_2   © Springer-Verlag Berlin Heidelberg 2006

lines." Weyl defined the terms *affine* and *linear* to be synonymous ([115], end of p. 21). However, this is no longer so in current usage. In current usage, linear transformations are *homogeneous*, and the group of linear transformations of $\mathbb{R}^n$ is the group $GL(n, \mathbb{R})$. The group of affine transformations of $\mathbb{R}^n$ is a semidirect product of T_n, the group of translations of $\mathbb{R}^n$, and the group $GL(n, \mathbb{R})$. In the physics literature, this group is sometimes called the *inhomogeneous general linear group*, and denoted by $IGL(n, \mathbb{R})$. We shall use the term "linear" in its current mathematical sense. The term "affine" will be synonymous with "inhomogeneous linear". The strictly inhomogeneous part of an affine transformation will be called a "translation". The term *affine connection*, still in use, owes its origin to the link between the term "affine" and the translations of $\mathbb{R}^n$.

2. **The Euclidean Structure:** This is defined by the Euclidean metric

$$d(x, y) = \{(x_1 - y_1)^2 + \cdots + (x_n - y_n)^2\}^{1/2} , \tag{2.1}$$

where $x = (x_1, \ldots, x_n)$, $y = (y_1, \ldots, y_n)$ are points of $\mathbb{R}^n$. This metric defines a topology which is equivalent to the product topology that we have assumed, and defines an inner product via the law of cosines of trigonometry. The group which preserves this inner product structure is known as the *orthogonal group*, and is denoted by $O(n)$. It is the maximal compact subgroup of $GL(n, \mathbb{R})$. The inhomogeneous orthogonal group is known as the Euclidean group.

3. **The Minkowski Structure:** This is defined by the indefinite Minkowski form (called the Minkowski metric in physics)

$$s(x, y) = \{(x_0 - y_0)^2 - (x_1 - y_1)^2 - \cdots - (x_{n-1} - y_{n-1})^2\}^{1/2} . \tag{2.2}$$

We have chosen the signature $+ - \cdots -$ because, in field theories[2] on Minkowski space, positivity of the energy is an important consideration. In the literature on general relativity, the opposite signature, $+ \cdots + -$, is frequently used, but there is no unanimity on this question. This choice induces a positive metric on spacelike hypersurfaces in (special and) general relativity. Historically, at the beginning of the theory of relativity one often introduced the imaginary variable $x_n = ix_0$, in order to make the Minkowski form look like the Euclidean metric $x_1^2 + x_2^2 + \cdots + x_n^2$, which too led to the Minkowski signature $+ \cdots + -$. Lorentz transformations are linear transformations of $\mathbb{R}^n$ that leave the Minkowski form invariant. The group of Lorentz transformations, called the *Lorentz group*, is denoted by $O(1, n-1)$, or more often by L. The *inhomogeneous* Lorentz group is also known as the *Poincaré* group.

---

[2] In this monograph the term *field theory* will only mean a classical or quantum field theory in physics. We shall have no occasion to refer to the theory of fields which is a branch of mathematics.

2.2 Local Structures on Manifolds

The manifestly geometrical character of the general theory of relativity led to attempts at geometrizing the classical theory of the electromagnetic field, and at finding unified field theories; new questions were asked about the nature of space. Weyl was one of the first to realize that "space" was an interweaving of distinct mathematical structures and that these structures merited independent mathematical study. This study was central to his attempts at the geometrization of physics, which gave rise to the concept of *gauge invariance* that – albeit in a different form – was to become one of the dominant principles of physics in the second half of the 20th Century.

Weyl's interests were wide. In the preface to the first edition of *Raum–Zeit–Materie*, he wrote (quoted from [115]):

> Although... a series of introductions into the general theory of relativity has appeared, nevertheless a systematic presentation was lacking. I therefore considered to publish the following lectures which I gave in the Summer Term of 1917 at the *Eidgen. Technische Hochschule* in Zürich. At the same time it was my wish to present this great subject as an illustration of the intermingling of philosophical, mathematical, and physical thought, a study which is dear to my heart.

The reader who wishes to pursue this "intermingling of philosophical, mathematical and physical thought" could do no better than start with the recent volume *"Hermann Weyl's* Raum–Zeit–Materie *and a General Introduction to his Scientific Work"* [93], and proceed to Weyl himself (particularly [120]; see the last Section of this Introduction). We shall content ourselves with a brief account of the mathematical aspect of the subject from the current viewpoint; this may also facilitate the reading of the original sources.

2.2.1 Remark on Terminology

The initial development of Riemannian geometry was confined to Riemannian metrics $ds^2 = g_{ij}dx^i dx^j$ (we assume the Einstein summation convention) that were positive-definite. However, after the advent of the general theory of relativity it was realized that many of the key concepts remained meaningful, and the results valid (perhaps in a modified form), even without this restriction. Indeed, the restriction to positive-definite Riemannian metrics was not made in Eisenhart's book *Riemannian Geometry* ([35]; see the Preface). It follows that (in developing Riemannian geometry through the tensor calculus) it is not really necessary to distinguish, terminologically, between Riemannian and pseudo-Riemannian metrics. However, the theorem that a finite-dimensional differentiable manifold[3] always admits a Riemannian metric holds only if

---

[3] Henceforth the term 'differentiable manifold' will mean a *finite-dimensional* manifold, unless stated otherwise; infinite-dimensionality will always be made explicit. See also Sect. 2.2.2, and footnote 3 on p. 172.

the Riemannian metric is understood to be positive-definite (see below), and therefore we shall stay with this terminological distinction.

2.2.2 Differentiable Manifolds and Geometry

If (as is generally the case) one assumes that space-time has a differentiable structure, the appropriate mathematical tools for its study would be differential geometry for local problems and differential topology for global problems. Both start by making precise the notion of a differentiable manifold.

We recall some notations and terminology. A topological space which is locally homeomorphic to $\mathbb{R}^n$ for some n is called an n-dimensional topological manifold, or a $C^{(0)}$-manifold. A $C^{(k)}$-differentiable manifold, $k = 1, 2, \ldots$ is one that carries a k-fold differentiable structure. A $C^{(\infty)}$-manifold is also called a smooth manifold. We remind the reader of the following facts:

1. There exist topological manifolds that admit of no differentiable structure [58].
2. Every $C^{(k)}$-structure on a manifold ($k > 0$) admits a compatible $C^{(\infty)}$-structure (see, for example, [50]). For most practical purposes, therefore, it may be assumed that a differentiable manifold is smooth.[4]
3. Every differentiable manifold admits a (positive-definite) Riemannian metric (see Appendix B, or [75]).

As a consequence of the second, there is no real loss of generality, in the differential category, in restricting attention to smooth manifolds. As a matter of historical interest we remark that the definition of a differentiable manifold by means of overlapping coordinate patches was first given by Weyl in 1912 in his book *Die Idee der Riemannschen Fläche* [112]. However, the subject became established as an independent discipline only much later, after the papers of Whitney, the first of which [123] appeared in 1936.

The tabula rasa of today's geometer is the differentiable manifold; like the vacuum state of today's physicist, it already carries a lot of information.

We shall now describe two further geometrical structures that were introduced by Weyl; these structures are *local*:

2.2.3 The Conformal Structure

The local version of the Minkowski structure is called a *Lorentz structure*, and is most often expressed in the physics literature as a family of (local) coordinate transformations that leave the pseudo-Riemannian metric (called simply the *metric* in general relativity)

$$ds^2 = g_{\mu\nu}(x)dx^\mu dx^\nu \tag{2.3}$$

---

[4] The differentiable structures in this monograph will always be assumed smooth.

invariant at each x. Passing to the tangent bundle[5] $T(M)$, one finds that they define a family of coordinate transformations on M with values in $O(1, n-1)$, the Lorentz group. In his attempt to unify general relativity and electrodynamics, Weyl [113] was led to consider a more general class of transformations that do *not* leave ds^2 invariant but rather multiply it by a local scaling factor, a "gauge":

$$ds^2 \to \Omega(x)\, ds^2 = \Omega(x)\, g_{\mu\nu} dx^\mu dx^\nu, \tag{2.4}$$

where $\Omega(x)$ is some smooth function of x. These transformations were called *conformal* by Weyl, because they preserved angles (see [35]). Additionally, while they did not leave ds^2 invariant, they clearly left the surfaces $ds^2 = 0$ invariant, i.e., they mapped light cones to light cones. Conformal transformations on a 4-dimensional Minkowski space form a 15-parameter group called *the* conformal group,[6] which is the Lorentz group extended by dilatations and the transformations of reciprocal radii x_i to x_i/x^2.[7]

It follows that conformal transformations on "flat" Minkowski space that do not leave $ds^2 \neq 0$ invariant have to be nonlinear. If the dimension of the Minkowski space is greater than 2, then there exist conformal transformations that send timelike differences to spacelike ones. This cannot happen in two dimensions, which is why conformal field theories are mostly studied on two-dimensional Minkowski space.

It was shown by Weyl that conformal transformations left the tensor

$$C^\lambda_{\mu\nu\sigma} = R^\lambda_{\mu\nu\sigma} + \frac{1}{n-2}\left(\delta^\lambda_\nu R_{\mu\sigma} - \delta^\lambda_\sigma R_{\mu\nu} + g_{\mu\sigma} R^\lambda_\mu - g_{\mu\nu} R^\lambda_\sigma\right)$$
$$+ \frac{R}{(n-1)(n-2)}\left(\delta^\lambda_\sigma g_{\mu\nu} - \delta^\lambda_\nu g_{\mu\sigma}\right) \tag{2.5}$$

invariant. In the above, $R^\lambda_{\mu\nu\sigma}$ is the Riemann tensor, $R_{\mu\sigma}$ is the Ricci tensor, and R is the scalar curvature. The tensor $C^\lambda_{\mu\nu\sigma}$ is known by various names: Weyl himself called it the conformal curvature tensor [35]. Current usage seems to favour the terms Weyl tensor [32] or the conformal tensor [110].

2.2.4 The Weyl Projective Structure

The notion of *geodesics* is defined on Riemannian and pseudo-Riemannian manifolds. By analogy with affine transformations on flat spaces, the study of transformations that map geodesics to geodesics is clearly of interest. These transformations were studied by Weyl, who showed that they are characterized by the fact that they leave the following tensor invariant:

---

[5] See Appendix B for the definitions of fibre bundles and G-structures.

[6] In the mathematical literature, the conformal group is generally defined on *Riemannian* manifolds of arbitrary dimension. See the Remark B.2.3, Appendix B.

[7] In quantum field theory, this map is usually taken with a minus sign: $z_i \to -z_i/z^2$, so that the complex forward tube is mapped into itself. See [126].

$$W^\lambda_{\mu\nu\sigma} = R^\lambda_{\mu\nu\sigma} + \frac{1}{n-1}(\delta^\lambda_\nu R_{\mu\sigma} - \delta^\lambda_k R_{\mu\nu}) \,. \tag{2.6}$$

He called the tensor (2.6) the *projective curvature tensor* [35], and the structure associated with it the *projective structure*. Note that, in a pseudo-Riemannian manifold, a projective structure does not necessarily map null geodesics to null geodesics, or light cones to light cones.[8]

In general relativity, timelike geodesics correspond to paths of freely falling particles of nonzero rest mass. Ehlers, Pirani and Schild [32] have studied the local structure of a manifold in which conformal and (Weyl) projective structures are defined locally.

2.3 Nature of the Present Work

In developing his "infinitesimal geometry" Weyl, like Felix Klein before him, shifted attention from the geometrical object itself to a family of transformations of the object. In Klein's case,[9] the transformations formed a group; in Weyl's case, they formed a Lie algebra[10] which left his conformal (or projective, as the case may be) curvature tensor invariant. The transformation group did not define a unique underlying space, a fact that was once regarded as a weakness of the Erlangen Programme. Weyl turned it into a strength, by considering Lie algebras that preserved only *some* of the structures on an underlying space. Conformal transformations did not preserve the full Lorentz structure, and Weyl projective transformations did not preserve the full affine structure. Passage to Lie algebras enabled Weyl to exploit the similarities – which would be called the differentiable structure today – to explore the differences.

From this perspective, it is possible to view the work presented in this monograph as follows:

1. First, the study of conformal structures as defined by Weyl in their "purest form", without reference to a notion of length, or indeed that of real numbers.[11] This, of course, will erase the differential structure on which Weyl based his analysis.
2. Second, the search for natural embeddings of these spaces – if any – into others that can carry a differential structure.

It is interesting to ask whether or not a similar programme can be carried out for the Weyl projective structure. The results of Ehlers, Pirani and

---

[8] Note that this projective structure is not related to the real projective spaces $\mathbb{R}P^n$, which are nonorientable manifolds.

[9] The reference here is to the *Erlanger Programm* of 1872. See, for instance, [7].

[10] The term *Lie algebra* was coined by Weyl himself.

[11] "The introduction of numbers as coordinates... is an act of violence...", Hermann Weyl [120], quoted in [50].

Schild[12] [32] suggest that the answer would be in the affirmative, but such an analysis has not yet been carried out.

The above remarks also explain our choice of title for this volume.

2.4 Weyl on the Geometry of Space-Time

The Fourth Edition of Weyl's *Raum–Zeit–Materie* was translated into English under the title *Space–Time–Matter* [115] in 1922. Weyl prepared a fifth German edition [116], with an expanded chapter on the General Theory of Relativity, which appeared in 1923. The physics behind the conformal and projective structures was explained in this edition (Chap. IV), but their "infinitesimal geometry" (which would be called differential geometry today) was not elaborated.

The mathematical arguments that led to Weyl's conformal (2.5) and projective (2.6) curvature tensors are given in full in Eisenhart's *Riemannian Geometry* [35]. Weyl's original account of the conformal curvature tensor was published in 1918, in [113]. The projective curvature tensor made its first appearance in 1921, in [114], which was probably Weyl's most detailed work on the subject. An account in English, entitled *On the foundations of general infinitesimal geometry* and written by Weyl himself, appeared in 1929 [119].

In the preface to the fifth edition of *Raum–Zeit–Materie*, Weyl mentioned his group-theoretical analysis of the subject, and directed the reader to his 1922 lectures (in Barcelona and Madrid) on the *Mathematische Analyse des Raumproblems* which was due to be published, in Spanish, by the Institut d'Estudis Catalans (Barcelona). He added that it would perhaps be published in German as well; it was, by Julius Springer in Berlin, in 1923 [117]. No English translation has appeared.

Weyl's philosophical reflections on space-time are never far from the physics and mathematics in *Raum–Zeit–Materie*. A version addressed chiefly to philosophers was published in 1926 [118]. A revised and augmented English translation was published by the Princeton University Press in 1949 [120].

Weyl's papers have been reprinted in four volumes in his *Gesammelte Abhandlungen* (Collected Papers). Papers [113] and [114] are to be found in Vol. II, and [119] in Vol. III.

In 1985, the hundredth anniversary of Weyl's birth, the university of Kiel organized an "International Hermann Weyl Congress" on *Exact Sciences and their Philosophical Foundations.* Twenty lectures given at this congress, covering the entire range of Weyl's work on mathematics, physics and the philosophy of science, edited by Deppert, Hübner, Oberschelp and Weidemann, were published in [21].

In 1988, ETH Zürich organized a Weyl Centenary celebration. The Weyl Centenary volume, [18], contains articles by C. N. Yang (*Hermann Weyl's*

---

[12] This work will be discussed in Chap. 10.

Contribution to Physics), Roger Penrose (*Hermann Weyl, Space-Time and Conformal Geometry*) and Armand Borel (*Hermann Weyl and Lie Groups*). On this occasion, the ETH also published a selection of Weyl's papers in one volume, entitled *Selecta Hermann Weyl* [122]. Paper [114] is reprinted in this volume.

In 1994, the *Deutsche Mathematiker Vereinigung* organized a seminar on Weyl's "contributions to the rise of general relativity and unified field theories". The book *Hermann Weyl's* Raum-Zeit-Materie *and a General Introduction to his Scientific Work* [93], published in 2000, grew out of this seminar. It consists of two parts. The first is devoted to "Historical Aspects of *Raum–Zeit-Materie*", with contributions by S. Sigurdsson, E. Scholtz, H. Goenner and N. Straumann. The second part, by R. Coleman and H. Korté, is on "Hermann Weyl: Mathematician, Physicist, Philosopher". The volume is aimed at mathematicians, physicists and historians and philosophers of science, and has an extensive bibliography.

3

Light Rays and Light Cones

In this chapter we shall embark upon an axiomatization of Einstein-Weyl causality on a set M consisting of points in the sense of Euclidean geometry.[1] Our aim is to investigate the *mathematical* consequences of Einstein-Weyl causality at the local level, and we therefore assume that M carries no pre-defined mathematical structure. Causal structures have been investigated by other authors, but mainly on spaces which already carried other structures such as a manifold structure (the work of Kronheimer and Penrose is a notable exception). Two of these will be discussed in some detail in Chap. 10.

Initially, the only proof techniques at our disposal are the basic ones: by explicit construction, or by contradiction. As a result, the argument may, at times, seem burdensome. We shall try to mitigate this by breaking up long proofs into smaller propositions, and by providing examples in two and three dimensional Euclidean space which can be drawn or visualized.

3.1 Light Rays and Order

There are several types or order on Minkowski space. There is, for example, the partial timelike order which corresponds to the physical constraint that a massive particle cannot attain the velocity of light. There is the partial order defined by light cones, which identifies the velocity of light as the fastest possible signal velocity; and then there is the order on the space-time path of an individual object or particle, which is a total order. Each is an essential component of Einstein–Weyl causality.

The aim of a mathematical "axiomatization" of Einstein–Weyl causality should presumably be to endow a point-set with the above properties via the weakest possible set of axioms. Clearly, as the structure is so rich, there are many possible points of departure. The one that we have chosen – after a

---

[1] By implication, we are assuming that the notion of a geometrical point makes physical sense, an assumption that should be made explicit.

H.-J. Borchers and R.N. Sen: *Mathematical Implications of Einstein–Weyl Causality*,
Lect. Notes Phys. **709**, 15–30 (2006)
DOI 10.1007/3-540-37681-X_3   © Springer-Verlag Berlin Heidelberg 2006

good deal of experimentation – is the total order on the space-time paths of light rays *in vacuo*.[2] In future, we shall use the term *light rays* as a shorthand for these paths, and the shorthand *causal order* to denote the collection of orders – partial or total – that reflect the physical principle of Einstein–Weyl causality.

Remarks 3.1.1 (The terminology of order) The terminology of order is not uniform, even in standard textbooks. For example, Kelley [57] regards order to be synonymous with partial order, and defines it to be a relation that is transitive. Dugundji [30], Munkres [77] and Willard [130] define partial order to be a relation that is reflexive, antisymmetric[3] and transitive. Munkres, additionally, defines strict partial order to be a relation that is transitive but nonreflexive; according to his definition, a strict partial order is not a partial order. Finally, in Kelly's definition ([57], p. 14), linear order is antisymmetric, but not necessarily reflexive. As far as we know, this lack of uniformity does not cause any confusion.

We shall use the term *order* in the sense of Kelley; that is, we shall assume neither reflexivity nor antisymmetry. When an order relation satisfies one or both of these conditions, we shall state it explicitly. Finally, we shall use the term *total order* for an order that satisfies the comparability condition, i.e., if the order $<$ is defined on X, $x, y \in X$, $x \neq y$, then either $x < y$ or $y < x$.

3.1.1 Light Rays and the Order Axiom

Thus, the fundamental objects in our scheme will be:

i) A nonempty set of *points M*.

ii) A distinguished family of subsets of M, called *light rays*.

iii) A total order $<^l$ (equivalently, $^l>$) on every light ray.

Points of M will be denoted by lower-case Latin letters. Light rays will be denoted by the letter l. l_x will denote a light ray through the point x, $l_{x,y}$ a ray through x and y, etc. Distinct rays will be distinguished by superscripts, thus l, l', l_x^1, l_x^2, etc. The statements $x <^l y$ and $y {}^l> x$ (read: x precedes y, or y follows x) will be identical. The notation $x <^{ll} y$ will mean that $x <^l y$ and $x \neq y$. The statement "x and y are joined by a light ray" will be abbreviated "$\lambda(x, y)$", and its negation (no light ray passes through both x and y) by "$\sim \lambda(x, y)$". These notations will be used in what follows without further comment.

A light ray will be assumed to satisfy the following:

---

[2] In Minkowski space these paths are straight lines with $s(x, y) = 0$, $x, y \in l$, where $s(x, y)$, the Minkowski form, is given by (2.2). This picture is abstracted from geometrical optics in the absence of dispersion.

[3] A relation R is called *antisymmetric* if aRb and bRa imply $a = b$.

Axiom 3.1.2 (The order axiom)

a) *If $x, y \in l$ and $x \neq y$, then either $x <^l y$ or $y <^l x$; if $x <^l y$ and $y <^l x$, then $x = y$.*

b) *If $x, z \in l$, $x <^{ll} z$, then $\exists y \in l$ such that $x <^{ll} y <^{ll} z$.*

c) *If $y \in l$, then $\exists x, z \in l$ such that $x <^{ll} y <^{ll} z$.*

d) *If $x, y \in l^1 \cap l^2$, then $x <^{l^1} y \Leftrightarrow x <^{l^2} y$.* □

(We shall use the symbol □ to denote the end of the statement of an axiom, a definition or a group of related definitions. See, however, footnote 4 on p. 18.)

Here condition a) merely states that the order $<^l$ is reflexive and antisymmetric. Condition b) states that between any two distinct points on a light ray there exists a third; it will, for this reason, be referred to as the *density axiom*. Condition c) states that light rays do not have end-points (end-points or singularities are admissible if they are not considered as parts of the space); and d) states that if the intersection of two light rays contains two distinct points, then these two points are similarly ordered with respect to the two rays; see Fig. 3.1. In the figure, the arrowheads denote the "direction of travel" of the light rays, and this part of the axiom may be thought of as saying that light rays do not travel "backward in time". It is a consistency condition that does not have to be imposed on Minkowski space.

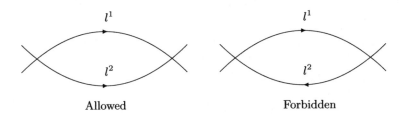

l^1 l^1

l^2 l^2

Allowed   Forbidden

Fig. 3.1. Illustrating the Order Axiom, Part d)

Our first problem is to extend the order on light rays to a partial order on all of M. This would hardly be possible if M were to consist of separate nontrivial pieces. However, we have not yet defined the term "separate" in the present context, and we proceed to fill this gap. We shall do this in two steps. In the first step, we shall ensure that if M contains a part of a light ray, then it contains the entire light ray. In the second step, we shall ensure that any point of M can be reached from any other point by a path consisting entirely of segments of light rays. It would be convenient to define two new concepts.

3.1.2 l-Completeness and l-Connectedness

We begin with the following definition:

Definition 3.1.3 A subset N of M will be called l-*complete* if $x \in N, l \ni x \Rightarrow l \subset N$. □

In words, if the subset N contains any part of a light ray, it contains the whole ray.

We shall now show that excising an l-complete subset from an l-complete set leaves an l-complete set, and that the property of l-completeness is stable under unions and intersections. Formally:

Lemma 3.1.4 *Let $\{N^\alpha\}_{\alpha \in A}$ be an indexed family of l-complete subsets of M. Then*

a) $M \setminus N^\alpha$ *is l-complete.*

b) $\bigcap_{\alpha \in A} N^\alpha$ *is l-complete.*

c) $\bigcup_{\alpha \in A} N^\alpha$ *is l-complete.*

Proof: The proofs are straightforward:

a) Let $x \in M \setminus N^\alpha, l_x \ni x$. Then $x \notin N^\alpha$, and therefore, by the l-completeness of N^α, $l_x \cap N^\alpha = \emptyset$. Hence $y \in l_x \Rightarrow y \in M \setminus N^\alpha$, and therefore $l_x \subset M \setminus N^\alpha$.

b) $x \in \bigcap_{\alpha \in A} N^\alpha \Rightarrow x \in N^\alpha \; \forall \alpha \in A$, therefore $l_x \subset N^\alpha$ for any l through x and any $\alpha \in A$, hence $l_x \subset \bigcap_{\alpha \in A} N^\alpha$.

c) $\bigcap_{\alpha \in A} (M \setminus N^\alpha) = M \setminus (\bigcup_{\alpha \in A} N^\alpha)$, and therefore

$$\bigcup_{\alpha \in A} N^\alpha = M \setminus \bigcap_{\alpha \in A} (M \setminus N^\alpha).$$

The result now follows from a) and b). ■

(We shall use the symbol ■ to denote "end of proof". On occasion, when the proof precedes the statement of the result, the symbol will be placed at the end of the statement.[4])

Construction of l-Complete Sets. Given any subset W of M, we would like to construct the "smallest" l-complete subset of M that contains W. This construction is as follows. Let A be any well-ordered set without largest element. Denote the smallest member of A by 0, and the successor of $\alpha \in A$ by $\alpha + 1$. Let

---

[4] The symbols □ and ■ will not be used for definitions or results that are part of the standard literature, e.g., Theorem 6.1.4 or the definitions and theorems in Appendix A.

$$K_W^0 \equiv W \subset M, \, W \neq \emptyset, \text{ otherwise arbitrary}, \tag{3.1}$$

$$K_W^1 \equiv \{y \, ; \, y \in l_z, z \in K_W^0\} \setminus K_W^0 . \tag{3.2}$$

In words, K_W^1 consists of all points on all light rays that pass through points of W, minus the set $W = K_W^0$ itself. Now define inductively

$$K_W^{\alpha+1} \equiv \{y \, ; \, \lambda(y, z), \, z \in K_W^\alpha\} \setminus \left\{ \bigcup_{\beta \leq \alpha} K_W^\beta \right\}, \, \alpha > 0 . \tag{3.3}$$

Note that $x \in K_W^{\alpha+1} \Rightarrow x \notin K_W^\alpha$. Finally, define

$$K_W^A \equiv \bigcup_{\alpha \in A} K_W^\alpha . \tag{3.4}$$

By construction, $x \in K_W^A \Rightarrow l_x \in K_W^A$ for every l through x, that is, K_W^A is l-complete.

The reader may be puzzled by condition (3.3) of the above construction, which ensures that the intersections $K_W^\alpha \cap K_W^{\alpha+1}$ are empty; it is clearly not needed for the purpose of defining the union K_W^A. This condition, together with (3.4), is designed to facilitate the constructive proof of Theorem 3.1.5 below.

In particular,[5] taking $A = \mathbb{N}$, we find that the subset $K_W^\mathbb{N}$ of M is l-complete. Define now the intersection of all l-complete sets containing W:

$$N_W \equiv \bigcap_{\substack{W \subset N, \\ N \text{ is } l\text{-complete}}} N \tag{3.5}$$

By Lemma 3.1.4, N_W is l-complete, and therefore

$$N_W \subset K_W^\mathbb{N} . \tag{3.6}$$

However, we have:

Theorem 3.1.5

$$N_W = K_W^\mathbb{N} .$$

Proof, abstract: In view of (3.6), it remains to prove that $N_W \supset K_W^\mathbb{N}$. For this, it suffices to observe that $\mathbb{N}$ is contained in every well-ordered set without largest element. ∎

This result, although simple, is of considerable practical significance. The constructive proof given below provides a foretaste of this:

---

[5] We shall use the symbol $\mathbb{N}$ to denote the set of nonnegative integers; for historical reasons, it is often used in the literature to denote the set of positive integers. We shall denote the latter set by $\mathbb{N}^+$.

Proof, constructive: If $a_0 \in K_W^{\mathbb{N}}$, then $\exists n \in \mathbb{N}$ such that

$$a_0 \in K_W^{n-1}, \quad a_0 \notin K_W^n .$$

Next, $b_j \in K_W^{n-j} \Rightarrow \exists b_{j+1} \in K_W^{n-j-1}$ such that $\lambda(b_j, b_{j+1})$ but $\sim\lambda(b_j, b_{j+2})$. Therefore there exists a set of points $\{a_0, a_1, \ldots, a_n\}$ such that

$$a_i \in K_W^{n-i}; \ \lambda(a_i, a_{i+1}), \sim\lambda(a_i, a_{i+2}) \text{ for } i = 1, 2, \ldots, n-1 \text{ and } a_n \in W .$$

Now $a_n \in W, \lambda(a_{n-1}, a_n)$ imply that a_{n-1} belongs to any l-complete set which contains K_W^0, and therefore $a_{n-1} \in N_W$. Repeating the same argument (with K_W^1 replacing K_W^0), we conclude that $a_{n-1} \in N_W \Rightarrow a_{n-2} \in N_W$, and so on, until finally $a_0 \in N_W$. Thus $K_W^{\mathbb{N}} \subset N_W$.

∎

Corollary 3.1.6 *If $y \in N_x = K_{\{x\}}^{\mathbb{N}}$, then there exists a positive integer n and a set of points $\{x_0, x_1, \ldots, x_n\}$ in M, with $x_0 \equiv x$ and $x_n \equiv y$, such that $\lambda(x_i, x_{i+1})$ and $\sim\lambda(x_{i-1}, x_{i+1})$ for $i = 1, 2, \ldots, n-1$.*

(**Remark:** The condition $\sim\lambda(x_{i-1}, x_{i+1})$ may later be replaced by a different condition, which, however, can be formulated only after the introduction of the order topology; see also Lemma 4.2.18.)

Proof: Specialize Theorem 3.1.5 to the case $W = \{x\}$, and take $x_i = a_{n-i}$, $i = 0, 1, \ldots, n$, $x_0 = x$, $x_n = y$.

∎

Corollary 3.1.7

$$y \in N_x \Leftrightarrow x \in N_y .$$

Proof: The subset $\{x_0, x_1, \ldots, x_n\}$ (in reversed order) is equally a subset of the set N_y.

∎

Theorem 3.1.8 *The relation $x \sim y$ iff $N_x = N_y$ is an equivalence relation.*

Proof: Follows from the properties of the equality sign in $N_x = N_y$.

∎

These equivalences classes are the "connected" pieces that we have been looking for. They are important enough to be given a name.

Definition 3.1.9 M will be called l-*connected* iff $M = N_x \ \forall x \in M$ (i.e. if M consists of a single equivalence class). □

The condition of l-connectedness rules out fibrations in which a light ray stays inside a fibre.

From now on we shall make the following nontriviality assumptions without further comment:

Assumptions 3.1.10 (Nontriviality assumptions)

a) *M is nonempty, and does not consist of a single point.*
b) *M does not consist of a single light ray.*
c) *M is l-connected.*

Definition 3.1.11 (Open and closed segments)

$$l(a,b) \equiv \{x \; ; \; x,a,b \in l, \; a <^{ll} x <^{ll} b\} \, ,$$

$$l[a,b] \equiv \{x \; ; \; x,a,b \in l, \; a <^{l} x <^{l} b\} \, . \qquad \square$$

$l(a,b)$ and $l[a,b]$ will be called open and closed *segments* (*not* intervals!) of light rays.

A concatenation of closed right ray segments will be called an *l-polygon*. An *l*-polygon need not be a closed figure.

Corollary 3.1.6 may be stated in words as follows: *Any two points of an l-connected space N can be joined by an l-polygon.*

3.1.3 The Identification Axiom

The class of spaces satisfying the order Axiom 3.1.2 is too large for our purposes. For example, nothing in the order axiom rules out the possibility that two light rays meet, "travel" as one, but eventually split and resume their separate identities. However, we do *not* want this degree of generality, and therefore adopt the following axiom:

Axiom 3.1.12 (The identification axiom) *If l and l′ are distinct light rays and $a \in S \equiv l \cap l′$, then there exist $p, q \in l$ such that $p <^{ll} a <^{ll} q$, and $l(p,q) \cap S = \{a\}$. Similarly for l′.* $\qquad \square$

This axiom ensures that the intersection of two light rays contains no "point of accumulation"; however, it allows gravitational lenses to exist (see Fig. 3.2), and, in that sense, is grounded in physical reality.

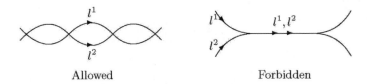

Allowed   Forbidden

Fig. 3.2. Illustrating the Identification Axiom

Example 3.1.13 Let M be the cylinder $S^1 \times \mathbb{R}$, with base S^1 placed horizontally. Let the light rays through any point be two lines each making an angle of $\pi/4$ with the vertical at that point. These two rays intersect infinitely many times. This example fulfills the order and identification axioms.

Example 3.1.14 This example consists of the one-sheeted hyperboloid (de Sitter space[6]), pictured so that the circular sections are horizontal. The light rays are the families of generators. Therefore two light rays do not intersect more than once.

The cylinder and the one-sheeted hyperboloid are topologically indistinguishable. Clearly, their order structures are very different, and, equally clearly, this difference appears at the global and not at the local level. These examples show that the order structure may be able to distinguish between topologically identical structures.

3.2 Construction of Cones

We shall now extend the total order on the light rays to a partial order on all of M. We shall do this by identifying "light cones" on M and determining the "interiors" and "boundaries" of these cones. For the moment, we are using the latter terms in an intuitive sense – no topology has yet been placed on M. We begin with the following definition:

Definition 3.2.1 A subset $W \subset M$ will be called *increasing* (respectively *decreasing*) if

$$x \in W, \, y \,{}^l\!> x \Rightarrow y \in W$$

$$(\text{respectively } x \in W, \, y <^l x \Rightarrow y \in W) \,. \qquad \square$$

Increasing and decreasing subsets can be constructed by an inductive process similar to the construction of l-complete sets in the previous section. We are interested in the smallest increasing (respectively decreasing) subset containing a given point x: from Def. 3.2.1 one sees that the property "increasing" (respectively "decreasing") is stable under set-theoretic intersections.

Using increasing and decreasing subsets, we define *cones* as follows:

Definition 3.2.2

$$C_x^+ \equiv \bigcap_{\substack{W \ni x \\ W \text{ increasing}}} W$$

$$C_x^- \equiv \bigcap_{\substack{W \ni x \\ W \text{ decreasing}}} W$$

and

---

[6] Let M be a locally Minkowski space of constant (Gaussian) curvature K, calculated with the pseudo-Riemannian metric. If $K = 0$, then M is Minkowski space itself. If $K < 0$, then M is called a *de Sitter* space; if $K > 0$, then M is called an *anti-de Sitter* space.

$$C_x \equiv C_x^+ \bigcup C_x^- .$$

□

C_x^+ will be called the *forward*, or *future* cone at x. Likewise, C_x^- will be called the *backward*, or *past* cone at x. The union C_x will be called the *cone* at x.

The following lemma will be the key technical tool for exploiting the properties of forward and backward cones:

Theorem 3.2.3 (The polygon lemma) *If $y \in C_x^+$, then there exist points $x_0, x_1, \ldots, x_{n-1}, x_n \in C_x^+$ such that $x_i <^{ll} x_{i+1}$ for $i = 0, 1, \ldots, n-1$ and $\sim \lambda(x_i, x_{i+2})$ for $i = 0, 1, \ldots, n-2$, with $x_0 = x$ and $x_n = y$. Similarly for $y \in C_x^-$ (see Fig. 3.3).*

(**Note:** The remark after Corollary 3.1.6 is valid also here.)

Proof: Let

$$W_x^0 \equiv \{x\} ,$$

$$W_x^{k+1} \equiv \Big\{ y; \; y \, ^l\!> z, \; z \in W_x^k \Big\} \setminus \Big\{ \bigcup_{\mathbb{N} \ni j \leq k} W_x^j \Big\}, \quad k \in \mathbb{N} .$$

Finally, let

$$W_x^+ \equiv \bigcup_{k \in \mathbb{N}} W_x^k .$$

Clearly, W_x^+ is an increasing set, and if $y \in W_x^+$ then there exist points $x_0, x_1, \ldots, x_n = y \in W_x^+$ such that $x_i <^{ll} x_{i+1}$ and $\sim \lambda(x_i, x_{i+2})$ for $i = 0, 1, \ldots, n-2$. It is therefore enough to prove that $W_x^+ = C_x^+$. From the definition of C_x^+, $W_x^+ \supset C_x^+$. The proof that $a \in W_x^+ \Rightarrow a \in C_x^+$, i.e. $W_x^+ \subset C_x^+$, is similar to the proof of the corresponding assertion in Theorem 3.1.5.

A similar proof holds for $y \in C_x^-$. ■

Let $\{x_0, x_1, \ldots, x_n\}$ be a finite set of points satisfying the conditions

$$x_i <^{ll} x_{i+1} \quad \text{and} \quad \sim \lambda(x_i, x_{i+2}) \tag{3.7}$$

for $i = 0, 1, \ldots, n-1$ and $i = 0, 1, \ldots, n-2$ respectively. The concatenation of the light-ray segments $l[x_k, x_{k+1}]$, $k = 0, 1, \ldots, n-1$ will be called an *ascending l-polygon* from x_0 to x_n, or a *descending l-polygon* from x_n to x_0. When there is no possibility of misunderstanding, the "*l-*" in the phrases above will be omitted. Corollary 3.1.6 and Theorem 3.2.3 will both be called the *polygon lemma*. The difference between the two will become apparent after the cone Axiom 3.2.8 is formulated (and which would imply that the cones are proper); Theorem 3.2.3 will then be seen to be restricted to ascending or descending polygons, which cannot exit from the (forward or backward) cone. Since the restriction to ascending or descending polygons will always have to be made explicit, the precise result being referred to will always be clear from the context.

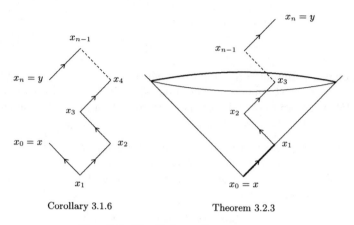

Corollary 3.1.6   Theorem 3.2.3

Fig. 3.3. The Polygon Lemma

Notations 3.2.4

The l-polygon determined by the set of points $\{x_0, x_1, \ldots, x_n\}$ satisfying the conditions (3.7) will be denoted by $P(x_0, x_1, \ldots, x_n)$.

Corollary 3.2.5

$$y \in C_x^+ \Leftrightarrow x \in C_y^- .$$

Proof: Consists of the observation that an ascending polygon from x to y is equally a descending polygon from y to x.   ∎

The polygon lemma says that if x and y are two distinct points in an l-connected space, then y can be reached from x by traversing a *finite number* of light-ray segments. This, it should be remembered, is a consequence of our definition of N_W (construction of l-complete sets, (3.5)) as the "smallest" l-complete set containing W. It would have been possible, at that stage, to define N_W as the intersection of *all* l-complete sets which are obtained by repeating the process (of joining with light rays) a *transfinite* (corresponding to a given cardinality) number of times, which would have led to transfinite l-polygons. However, we are eschewing this degree of generality.

We are now ready to extend the definition of order.

Definition 3.2.6 Define

$$x < y \ (\text{or } y > x) \quad \text{iff} \quad y \in C_x^+ ;$$

equivalently,

$$x < y \ (\text{or } y > x) \quad \text{iff} \quad x \in C_y^- . \qquad \square$$

Observe that if $\lambda(x, y)$ then $x < y \Rightarrow x <^l y$.

The basic result follows quickly:

Theorem 3.2.7 $<$ (or $>$) defines a reflexive partial order on M.

Proof: By definition, $x \in C_x^+$, and therefore $x < x$. To prove transitivity, let $x < y$, $y < z$. Then there is an ascending polygon from x to y, and one from y to z. The concatenation of the two is an ascending polygon from x to z, hence $x < z$. ∎

However, the order relation "$<$" is not antisymmetric. Let us begin by considering the physics. Owing to parts a) and c) of the order axiom (Axiom 3.1.2), a light ray cannot form a closed loop. But there is nothing to prevent the existence of closed timelike curves[7] and (physical) pathologies that follow from it, as shown by Example 3.2.9, which satisfies the order and identification axioms. We need a new axiom to eliminate these pathologies:

Axiom 3.2.8 (The cone axiom)

$$C_x^+ \bigcap C_x^- = \{x\} \quad \forall \, x \in M \qquad \square$$

Example 3.2.9 This example consists of the two-dimensional anti-de Sitter space. Like the two-dimensional de Sitter space (Example 3.1.14), this is also a one-sheeted hyperboloid in which the light rays are the generators. However, interiors and exteriors of light cones are interchanged in the de Sitter and the anti-de Sitter spaces. Consequently, the anti-de Sitter space admits closed timelike curves, which the de Sitter space does not. The configuration is shown in Fig. 3.4.[8]

This example satisfies the order axiom and the identification axiom, but violates the cone axiom.

With the addition of the cone axiom, we have:

Proposition 3.2.10 In a space M in which the cone axiom is satisfied, the order "$<$" is antisymmetric.

Proof: Suppose that $x < y$ and $y < x$, i.e., $y \in C_x^+$ and $x \in C_y^+$. By Corollary 3.2.5, $x \in C_y^+ \Rightarrow y \in C_x^-$, and therefore $y \in C_x^+ \cap C_x^-$. Therefore, by the cone axiom, $y = x$. ∎

---

[7] The precise definition of timelike curves is given in Def. 8.1.1, p. 103.

[8] We have chosen the anti-de Sitter space to illustrate the violation of the cone axiom because this space supports the Maldecena conjecture of string theory. In this case one has the anti-de Sitter space – conformal field theory (AdS-CFT) correspondence. For details, see the works of Rehren [87, 88] and Dütsch and Rehren [29], and references therein.

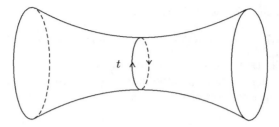

Fig. 3.4. This Anti-de Sitter Space Violates the Cone Axiom

Forward and backward cones have the following inclusion property:

Proposition 3.2.11

$$x < y \Leftrightarrow C_x^+ \supset C_y^+ \Leftrightarrow C_x^- \subset C_y^- .$$

Proof:

i) Let $z \in C_y^+$. Then $y < z$, therefore $x < y \Rightarrow x < z$, i.e., $z \in C_x^+$. Thus $x < y \Rightarrow C_x^+ \supset C_y^+$.

ii) If $C_x^+ \supset C_y^+$, then $y \in C_x^+$, i.e., $x < y$.

This establishes that $x < y \Leftrightarrow C_x^+ \supset C_y^+$. The equivalence $x < y \Leftrightarrow C_x^- \subset C_y^-$ is established similarly. ∎

Notations and Terminology 3.2.12

The intersection of a light ray l_x through x with the forward cone at x will be called the *forward ray* through x and denoted thus:

$$l_x^+ \equiv l_x \bigcap C_x^+ .$$

Similarly, the *backward ray* through x will be defined as $l_x^- \equiv l_x \cap C_x^-$. Furthermore, if $x <^{ll} y$, the notation $l_{x,y}^+$ will be used to denote the forward ray through x which passes through y. Similarly, $l_{y,x}^-$ will denote the backward ray from y which passes through x.

Henceforth we shall use the terms forward and backward rays, and the notations defined above, without comment.

3.2.1 Timelike Points

In a Minkowski space the boundary of the light cone through x consists of the set of all light rays through x. An interior (future) point y may be characterized by the fact that, given *any* light ray l_x through x, there exists a descending polygon from y to x that meets l_x at a point above[9] x. This latter property may be used to define "interior" points in our present setting. The "boundary" will then be defined as the residual set. All of this can be accomplished without defining a topology on M. However, to avoid confusion we shall use a modified topological terminology and notation until the topology has been defined.

It should be noted that the *finiteness* of the l-polygons plays an essential role in the following (see Theorem 3.2.3 and the comments following its proof).

Definition 3.2.13

a) A point $z \in C_x^+$ will be called a *timelike point* of C_x^+ if, for *any* ray $l_x \ni x$, $\exists y \in l_x^+$, $y \neq x$ such that $z \in C_y^+$. The set of all timelike points of C_x^+ will be denoted by τC_x^+, and called the τ-interior of C_x^+.

b) The set τC_x^- is defined similarly, i.e., by interchange of order. $\square$

Lemma 3.2.14 $y \in \tau C_x^+$ and $z > y \Rightarrow z \in \tau C_x^+$, and the same for reversed order.

Proof: For any l_x there exists, by Def. 3.2.13 above, a point $r \in l_x^+$ with $y \in \tau C_r^+$. Since $>$ defines a partial order (Theorem 3.2.7), the relation $z > y$ implies $z \in C_r^+$. Hence $z \in \tau C_r^+$. The result for reversed order follows in the same manner.   ∎

Lemma 3.2.14 implies that if a light ray enters $\tau C_x^\pm$ at its tip then it stays inside the light cone ("inner" light rays). This situation has to be excluded on physical grounds, and will be done via the local structure Axiom 4.2.1, which will be introduced in the next chapter.

Definition 3.2.15

$$\beta C_x^+ \equiv C_x^+ \setminus \tau C_x^+, \qquad \beta C_x^- \equiv C_x^- \setminus \tau C_x^-.$$

$\square$

We shall call τC_x^+ and βC_x^+ the τ-interior and the β-boundary of C_x^+ respectively (and similarly for C_x^-).

Remarks 3.2.16

1. Examples show that $\beta C^+{}_x$ and $\partial C^+{}_x$ may be different. A simple one is the two-dimensional Minkowski space with the strip $x \geq y$, $0 \leq y \leq 1$ excised, as shown in Fig. 3.5. The light ray l (dashed line) extending northeast from the point $(1,1)$ does not belong to the cone C_x^+ (shaded region). It is, however, part of the topological boundary ∂C_x^+ of C_x^+.

---

[9] To remove any possible misunderstanding, we remark that we shall use the terms "above x" (respectively "below x") to mean "in the forward (respectively backward) cone at x".

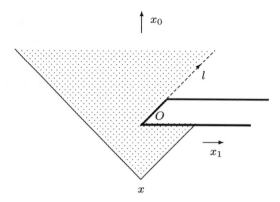

Fig. 3.5. Illustrating that βC_x^+ and ∂C_x^+ may be different

2. Note carefully that $y \in \beta C_x^+$, $y \neq x$ does *not* yet imply that y lies on a light ray through x; this desirable property will be established later using the convexity Axiom 4.2.1.d.

Definition 3.2.17

$$I[a, b] \equiv C_a^+ \bigcap C_b^- \ ,$$

$$I(a, b) \equiv \tau C_a^+ \bigcap \tau C_b^- \ .$$

The sets $I(a, b)$ and $I[a, b]$ will be called, respectively, *open* and *closed* order intervals. When the order topology is defined (Sect. 4.5), open and closed order intervals will turn out to be, respectively, open and closed sets, provided that there are no holes or cuts as in Fig. 3.5.   □

Remarks 3.2.18

It follows from the above definitions that (see also the Remark 4.2.2)

1. $I[a, a] = \{a\}$, $I(a, a) = \emptyset$.
2. If $a \not< b$, $a \neq b$ then $I[a, b] = \emptyset$.

In words, the second remark states that if a and b are mutually spacelike (see 3.2.24 at the end of this section for the terminology), or if $b \ll a$, then the order interval $I[a, b]$ is empty.

   In the rest of this section we shall establish a few results that we shall need in later chapters. These results may be intuitively obvious in Minkowski space, but, as the examples of Figs. 3.5 and 3.6 show, require proof in the present context.

Proposition 3.2.19 *Let $y > x$. Then the statements 1 and 2 (respectively 1' and 2') are equivalent:*

1. $y \in \tau C_x^+$;
2. $\beta C_x^+ \cap C_y^+ = \emptyset$.

1'. $x \in \tau C_y^-$;
2'. $\beta C_y^- \cap C_x^- = \emptyset$.

Proof:

Owing to symmetry, it suffices to establish one of the two equivalences:

$1 \Rightarrow 2$:
    Let $y \in \tau C_x^+$, and $z \in \beta C_x^+ \cap C_y^+$. Then $z > y$, hence $z \in \tau C_x^+$ by Lemma 3.2.14, a contradiction.

$2 \Rightarrow 1$:
    Since $y \in C_y^+$, $\beta C_x^+ \cap C_y^+ = \emptyset \Rightarrow y \notin \beta C_x^+$. But $y \in C_x^+$, therefore $y \in \tau C_x^+$. ∎

Although $y \in C_x^+ \Leftrightarrow x \in C_y^-$ (Corollary 3.2.5), it does *not* follow that

$$y \in \tau C_x^+ \Leftrightarrow x \in \tau C_y^- ; \tag{3.8}$$

statements about τC_x^+ and τC_y^- are unrelated (see Def. 3.2.21 and Fig. 3.6). However, if the equivalence (3.8) holds, then it can be expressed in several different ways:

Proposition 3.2.20 *The following statements are equivalent:*

1. $y \in \tau C_x^+ \Leftrightarrow x \in \tau C_y^-$;
2a. $y \in \tau C_x^+ \Rightarrow \forall l \ni y \; \exists z \in l_y^-, \; z \neq y$, *such that $z > x$; and*
2b. $x \in \tau C_y^- \Rightarrow \forall l \ni y \; \exists z \in l_x^+, \; z \neq x$, *such that $z < y$.*

Proof:

$1 \Rightarrow 2$:
    If $y \in \tau C_x^+ \Rightarrow x \in \tau C_y^-$, then (from the definition of τC_y^-) $\forall l \ni y, \exists z \in l_y^-$ such that $z > x$, i.e., 2a holds. Similarly, if $x \in \tau C_y^- \Rightarrow y \in \tau C_x^+$, it follows that 2b holds.

$2 \Rightarrow 1$:
    Assume that $y \in \tau C_x^+$, and $\forall l \ni y, \exists z \in l_y^-, z \neq y$, such that $z > x$. Then $x \in \tau C_y^-$. The same holds with the order reversed. ∎

Definition 3.2.21 An ordered space M will be said to have the *property S* (from symmetry) if

$$y \in \tau C_x^+ \Leftrightarrow x \in \tau C_y^- .$$

In this case M will be called an *S-space*, and we shall write

$$y \ll x \quad \text{iff} \quad y \in \tau C_x^+ . \qquad \square$$

Theorem 3.2.22 *In an S-space the relation $\ll$ defines a (nonreflexive, nonsymmetric) partial order.*

Proof: Transitivity is a consequence of Lemma 3.2.14.   ■

Remarks 3.2.23

1. If we define $y \gg x$ iff $x \in \tau C_y^-$, then, from Proposition 3.2.20, $x \ll y \Leftrightarrow y \gg x$.
2. Property S is not automatic, as shown by the example of Fig. 3.6. Two-dimensional Minkowski space is cut in two along the X-axis, leaving only the single point O to connect the half-planes. As the figure shows, descending l-polygons from b meet each of the two light rays from a at points above a. Therefore, by Def. 3.2.13, $b \in \tau C_a^+$. However, *no* ascending l-polygon from a can meet the ray l_b^{1-} at a point below b. Therefore $a \notin \tau C_b^-$.

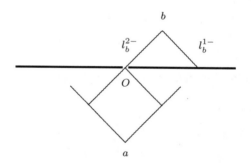

Fig. 3.6. Illustrating the Need for Property S

Terminology 3.2.24 In keeping with the usage in physics, we shall say that the points x, y are *spacelike, timelike or lightlike* with respect to each other according as i) $x \notin C_y$, ii) $x \gg y$ (or $x \ll y$) or iii) $\lambda(x, y)$. These three cases exhaust all possibilities.

4

Local Structure and Topology

4.1 Preliminary Remarks

There are many procedures for defining a topology on a space M. We select two for attention. In the first case, one may have a preferred family of functions defined on M. It may then be reasonable to consider the coarsest (weakest) topology on M which makes each of these functions continuous. In the second case, one may wish the space M to have strong homogeneity properties. Then M should be "glued together" from isomorphic copies of the same object.

In the present situation, we do not have a preferred family of functions on M. In view of Lebesgue's theorem that monotonic functions on the real line are differentiable almost everywhere, it would be tempting to try to define the topology via monotonic functions. But functions like the devil's staircase (see, for example, [22]) prevent this class from being really useful. Next, we may ask whether, on an ordered topological space, there is some sense in which order is continuous. If there is, is the coarsest topology which makes order continuous a useful one? Regarding the continuity of the order, let us recall the definition given for the ordered real line [57]: The order "<" is continuous if, given $x < y$, $x \neq y$, there exist neighbourhoods U of x and V of y such that $u < v$ whenever $u \in U$ and $v \in V$. And indeed, on $\mathbb{R}$ it turns out that the order topology (which coincides with the usual topology) is the coarsest topology which makes the order continuous.

Unfortunately, this definition fails to provide a physically acceptable topology in higher dimensions. To see this, consider two-dimensional Minkowski space, and two distinct point x and y on any light ray, with $x < y$. With the usual topology of this space, every neighbourhood of y contains points that are spacelike to x, and vice versa; the requirement that $u < v$ whenever $u \in U$ and $v \in V$, with U and V defined as above, cannot be met. In other words, with the usual topology, and with the above definition of the continuity of order, order is not continuous!

The concept of the order topology is, however, a useful one in ordered vector spaces, where it may be defined via order intervals. In our setting,

H.-J. Borchers and R.N. Sen: *Mathematical Implications of Einstein–Weyl Causality*,
Lect. Notes Phys. **709**, 31–50 (2006)
DOI 10.1007/3-540-37681-X_4

order intervals can have holes, and therefore not all of them will be suitable for our purposes. In Sect. 4.5 we shall show how the topology can be defined by "good" order intervals.

The precise manner that we have chosen to define the order topology is also reminiscent of the way in which differentiable manifolds are defined. That is, we make a distinction between the local and the global and demand that the order structure satisfy a strong homogeneity condition. The only way to determine a satisfactory condition is by trial and error. The choice that we have made, and its consequences, are presented in the following.

4.2 D-Sets and their Properties

We wish to embed every point in a neighbourhood which has all the desirable properties. The paradigm for this neighbourhood, which will be called a D-set (from the German *Durchschnittseigenschaft*), is the interior of a double cone (Int $C_x^+ \cap C_y^-$) in Minkowski space. An important property of a D-set is that it is an S-space (Def. 3.2.21). This, and a number of other properties that may not hold globally, hold locally–that is, in D-sets. The definition of a D-set is as follows.

Definition 4.2.1 A subset U of M will be called a D-*set* iff it fulfills the following conditions:

a) $x, y \in U \Rightarrow I[x, y] \subset U$ (Fig. 4.1(a)).

In words, U contains the entire (closed) order interval between any two of its points.[1] (If M were an ordered topological vector space, this would mean that U is order-convex. Although our ordered spaces are not generally linear spaces, we shall find it useful to employ the term *order-convex* in this sense.)

b) For every $x \in U$ and every $l \ni x$, there exist points $p, q \in l \cap U$ such that $p <^{ll} x <^{ll} q$ (Fig. 4.1(a)).

In words, the intersection of a light ray with a D-set does not have a minimal or a maximal point.

c) If $y \in U$, $r \in \tau C_y^- \cap U$ and $l_r \ni r$, then (Fig. 4.1(b), upper part)

$$l_r^+ \cap \{\beta C_y^- \setminus \{y\}\} \cap U \neq \emptyset,$$

and the same for reversed order.

In words, in a D-set, a forward ray from the τ-interior of a cone intersects its backward β-boundary below its vertex, and the same for reversed order.[2]

---

[1] Recall that if two points are spacelike to each other, then the order interval between them is, by definition (Def. 3.2.17), the empty set.

[2] It is worth remarking that this condition may not hold globally even in an ordered space which is homeomorphic with some $\mathbb{R}^n$ and in which two light rays do not

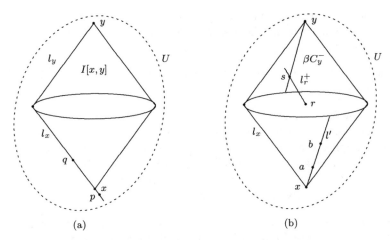

Fig. 4.1. Some defining properties of *D*-sets

d) If $x \in U$ and $l' \cap \beta C_x^+ \cap U$ contains two distinct points a, b, then (Fig. 4.1(b), lower part)

$$x \in l_{a,b}' \cap C_x^+ \cap U \subset \beta C_x^+ \cap U \ ,$$

and the same for reversed order.
In words, in a *D*-set U, a light ray that intersects the β-boundary of a forward cone at two distinct points lies wholly on the β-boundary of that forward cone (intersected with U), and passes through its vertex.

e) If $a, b \in U$ and $\lambda(a, b)$, then the ray $l_{a,b}$ is unique.
That is, two distinct light rays cannot intersect more than once in a *D*-set.

f) If $x \in U$, then there pass at least two distinct light rays through x.   □

Remark 4.2.2 (Addendum to Remarks 3.2.18)

One may add the following to the Remarks 3.2.18: If U is a *D*-set, $a, b \in U$ and $b \in \beta C_a^+$, then $I[a, b] = l[a, b]$ and $I(a, b) = \emptyset$.

Remark 4.2.3 (About the Figures)

In many of the diagrams, the boundary of the *D*-set under consideration is shown by a broken line enclosing an oval or a circular area. This is purely schematic, as disks and ovals in two-dimensional Minkowski space are *not D-sets*; they are not order-convex. Consider the disk K of radius $\sqrt{2}$ and centre $(x, y) = (0, 0)$ in $\mathbb{R}^2$. Denote the points $(-1, -1)$ and $(-1, 1)$ on its perimeter

---

intersect more than once. An example is provided by the universal cover of the anti-de Sitter space of Example 3.2.9. Coverings of ordered spaces are discussed in Sect. 8.5.

by a and b respectively. Then $\beta C_a^+ \cap \beta C_b^- = \{(-2,0),(0,0)\}$. A substantial part of $I[a,b]$ lies outside the disk K, which shows that its interior cannot be order-convex.

Observe that the empty set $\emptyset$ satisfies all of these conditions trivially, and is therefore a D-set.

Remark 4.2.4 Condition d) of Def. 4.2.1 will be called the *convexity axiom*. The reason for this terminology is as follows. In a linear space Condition 4.2.1.d) implies that the base of a light cone is absolutely convex, or equivalently that the faces of the cone are one-dimensional. (Recall that a closed convex set S is called absolutely convex if every straight line that meets ∂S at two distinct points also contains interior points of S.)

Remark 4.2.5 A D-set will be called *l-convex*, meaning that a light ray from its interior intersects its boundary[3] at exactly two points. However, it is not necessarily convex. Take a closed order interval in a D-set and drag it along an equatorial circle so that the doughnut-shaped object produced lies entirely in a D-set. This object is the closure of a D-set, but it is not a convex body (see Fig. 4.2; the shaded region is the "hole" in the doughnut).

Lemma 4.2.6 *Let U be a D-set and $I[a,b] \subset U$ a (closed) order interval. Let $x \in I(a,b)$. Then any light ray through x intersects $\beta I[a,b]$ at exactly two points, one lying on βC_a^+ and the other on βC_b^-.*

Proof: This follows immediately from the Definitions 4.2.1.c) and 4.2.1.d). ∎

The need for considering order intervals (open or closed) that lie entirely in D-sets will arise so frequently that it would be useful to give them a name:

Terminology 4.2.7 (D-intervals) An order interval $I(a,b)$ or $I[a,b]$ that lies entirely within a D-set will be called a D-interval.

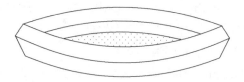

Fig. 4.2. A D-Set is not necessarily convex

We shall now establish the basic properties of D-sets. For the convenience of the reader, we shall group the results into classes, although the classes will

---

[3]   At this stage we cannot make the above remark into a formal definition, because the topology (and therefore the term boundary) has not yet been defined.

not be wholly disjoint. Initially, we shall allow ourselves a certain "abuse of notation", as explained in the following remark.

Remark 4.2.8 (Abuse of notation) Recall that the partial order $\ll$ is defined only on spaces that have the property S (see 3.2.21 and 3.2.22). In the following, we shall prove (in Theorem 4.2.15) that every D-set enjoys the property S. However, we shall, by abuse of notation, write $x \ll y$ instead of $x \in \tau C_y^-$ and $q \gg p$ instead of $q \in \tau C_p^+$ in D-sets for stating and proving the results 4.2.9–4.2.14 which precede the proof of Theorem 4.2.15. This is harmless, as property S is not invoked in the proofs of these results.

a) *D*-sets and timelike order

In Sect. 8.1 we shall see that, in a D-set, there exist subsets totally ordered by "$\ll$" that are homeomorphic to light ray segments. We begin by showing here that there are "enough" points in timelike order in a D-set.

Theorem 4.2.9 Let U be a nonempty D-set, and let $y \in U$. Then there exist points $x, z \in U$ such that $x \ll y \ll z$.

In words, every point in a D-set has a timelike predecessor and a timelike successor[4].

Proof: By condition c) of Def. 4.2.1, at least two light rays pass through y. Let l_y be one of them. Then, by Condition 4.2.1.b), there exists a point $p \in l_y^+ \cap U$, $y <^{ll} p$ (see Fig. 4.3). Then, by Condition 4.2.1.c), there is at least one more light ray l_p through p. Now, by Condition 4.2.1.b), there exists a point $z \in l_p^+ \cap U$, $p <^{ll} z$. Together, $y <^{ll} p$, $p <^{ll} z$ and $\sim\lambda(y,z)$ imply that $y < z$. Then either $y \ll z$ or $z \in \beta C_y^+$. If $z \in \beta C_y^+$ then, since $p \in \beta C_y^+$, it follows that $l_p \cap \beta C_y^+ \cap U$ contains two distinct points p and z. Therefore, from the convexity Axiom 4.2.1.d), it follows that $z \in l_p \cap C_y^+ \cap U$, i.e., the ray l_p passes through y, a contradiction. Therefore $y \ll z$.

The same argument, with order reversed, establishes the existence of a point x with the desired properties. ∎

The next result shows that, in a D-set, between any two distinct timelike points lies a third.

Lemma 4.2.10 Let U be a D-set, $x, z \in U$ with $x \ll z$. Then there exists $y \in U$ such that $x \ll y \ll z$.

Proof: We have to prove that $I(x,z)$ is nonempty. Let l_x^+ and l_z^- be light rays that intersect (such rays exist, by condition c) of Def. 4.2.1), and set $\{p\} = l_x^+ \cap l_z^- \subset U$. Pick a point $a \in l(x,p)$ and a point $b \in l(p,z)$ (see

---

[4] The terms *predecessor* and *successor* do not refer to immediate predecessors or immediate successors, as the sets we deal with are seldom well-ordered.

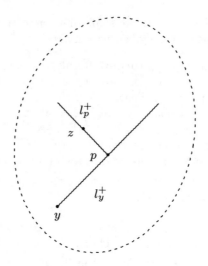

Fig. 4.3. Every point in a D-set has a timelike successor

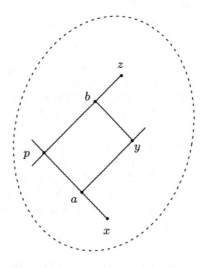

Fig. 4.4. Proving that $I(x, z) \subset U$ is nonempty

Fig. 4.4), and let l_b^- be a second backward ray from b. This ray intersects βC_a^+ at a point y, and $y \in I(x, z)$.   ■

The final result is an immediate corollary of Theorem 4.2.9:

Corollary 4.2.11 *Let U be a D-set, and let $x_0 \in U$. Then there exists an infinity of points $x_n \in U, n \in \mathbb{Z}$, such that $x_{n+1} \gg x_n$ for all $n \in \mathbb{Z}$.*

Proof: Consists of the repeated application of Theorem 4.2.9 in both forward and backward directions. ∎

b) Light rays from τ-interiors

Condition c) of Definition 4.2.1 states that, in a D-set, the intersection of a forward ray from the τ-interior with the β-boundary of the cone is nonempty. We prove below that this intersection consists of a single point:

Theorem 4.2.12 *Let U be a D-set, and let $x, y \in U$ with $y \gg x$. Let $l_y \ni y$ be a light ray through y. Then*

$$l_y \cap \beta C_x^+$$

consists of a single point, and the same with order reversed.

Proof: By Proposition 3.2.19, $y \gg x$ is equivalent to $\beta C_x^+ \cap C_y^+ = \emptyset$. But, by Def. 4.2.1.c), if $l_y \cap \beta C_x^+$ is nonempty then

$$l_y \cap \beta C_x^+ = l_y^- \cap \beta C_x^+ \subset I[x,y] \subset U .$$

Then, if $l_y^- \cap \beta C_x^+$ contained two distinct points, it would follow from Def. 4.2.1.d) that

$$l_y \cap C_x^+ \cap U \subset \beta C_x^+ ,$$

contradicting the assumption $y \gg x$. The same argument holds with order reversed. ∎

c) Incidence of light rays on cone boundaries

We now investigate the incidence relations[5] between light rays through the vertex of a cone and the β-boundary of the cone. We present two results that hold in D-sets: the first is that every point on the β-boundary of a cone is connected to its vertex by a light ray; the second is that every light ray through the vertex lies wholly on the β-boundary of the cone.

Lemma 4.2.13 *Let U be a D-set and let $x \in U$. Let $a \in \beta C_x^+ \cap U, a \neq x$. Then there exists a light ray $l_{x,a}$ such that $x, a \in l[x, a] \cap \beta C_x^+$ and $x <^{ll} a$.*

Proof: Let $P(x_0, \ldots, x_{n-1}, x_n)$ be any ascending l-polygon from $x = x_0$ to $a = x_n$. Suppose that $x_{n-1} \in \beta C_x^+$. Then, by the convexity Axiom 4.2.1.d), $x \in l_{x_{n-1},a}$, i.e., there exists a light ray $l_{x,a}$ through x and a, and $x_{n-1} \in l[x, a]$.

Suppose, next, that $x_{n-1} \notin \beta C_x^+$. Then $x_{n-1} \in \tau C_x^+$. Now $a = x_n > x_{n-1}$, and therefore it follows from Lemma 3.2.14 that $a \in \tau C_x^+$, a contradiction which proves the result. ∎

---

[5] Note that the convexity Axiom 4.2.1.d) itself is such a relation.

Lemma 4.2.14 *There are no "inner" light rays in a D-set* (see the remark after Lemma 3.2.14).

Proof: An inner light ray in C_x^+ is a light ray $l_{x,z}$ that connects x with a point $z \in \tau C_x^+$. By Lemma 4.2.12, a backward ray from z intersects βC_x^+ at exactly one point. By Axiom 4.2.1.c), this point cannot be the point x itself. ∎

d) The Property S

We now have the tools required to establish the following fundamental result:

Theorem 4.2.15 *Every D-set U has the property S.*

Proof: Let $x, y \in U$ with $y \gg x$ ($y \in \tau C_x^+$). Let $l_y \ni y$. Then, by Lemma 4.2.12, $l_y^- \cap \beta C_x^+$ is a unique point, say z. If $z \neq x$ it follows that $x \ll y$ ($x \in \tau C_y^-$). The case $z = x$ is ruled out by Lemma 4.2.14. ∎

e) New D-sets from old

There are plenty of D-sets. We shall establish this via two results; the first is that the intersection of two D-sets is a D-set, and the second is that an open order interval that is contained in a D-set is itself a D-set.

Proposition 4.2.16 *The intersection of two D-sets is a D-set.*

Proof: Let U_1, U_2 be D-sets. If $U_1 \cap U_2 = \emptyset$ then it is trivially a D-set. Assume now that $U_1 \cap U_2 \neq \emptyset$. Clearly, Condition 4.2.1.a) and conditions 4.2.1.c)–f) are stable under finite intersections. It remains to show that Condition 4.2.1.b) continues to hold.

Let $x \in U_1 \cap U_2$ and $l_x \ni x$. Then there exist points $p_i, q_i \in U_i$, $i = 1, 2$, such that $p_i, q_i \in l_x$, $p_i <^{ll} x <^{ll} q_i$. Since l_x is totally ordered it follows that for proper choice of indices a, b, m, n ($= 1$ or 2), $p_a > p_b$ and $q_m < q_n$. Then $p_b < x < q_n$, $p_b \neq x$, $q_n \neq x$, and $p_b, q_n \in U_1 \cap U_2$. ∎

The intersection property, Proposition 4.2.16, is the first step in defining a topology on M with the help of D-sets. This topology will be defined later and will be called the *order topology*. We shall then see that the topology induced on light rays by the order topology of M is the standard order topology on a totally ordered space. The validity of Condition 4.2.1.b) will then be seen to follow from the simple fact that the intersection of two open segments of a light ray is again an open segment of the ray.

Proposition 4.2.17 *Let U be a D-set, $x, y \in U$ and $y \gg x$. Then 1) $I(x, y)$ is a D-set; 2) $I(x, y)$ is l-connected.*

Proof of Part 1: Note that, by Lemma 4.2.10, $I(x, y)$ is nonempty. We shall verify that conditions a)–f) of Def. 4.2.1 hold for $I(x, y) \subset U$.

a) Let $p, q \in I(x, y)$. By Lemma 3.2.14, $C_p^+ \subset \tau C_x^+$ and $C_q^- \subset \tau C_y^-$. Hence $I[p, q] \subset I(x, y)$. This verifies Condition 4.2.1.a).

b) Take $p \in I(x, y)$ and $l \ni p$. Then by Lemma 4.2.12 the sets $l \cap \beta C_x^+ \equiv \{r\}$ and $l \cap \beta C_y^- \equiv \{s\}$ are singletons, differing from $\{p\}$. Therefore the open segment $l(r, s)$ belongs to $I(x, y)$, and Condition 4.2.1.b) is fulfilled by Axiom 3.1.2.c)–f). These are obviously true of order-convex subsets of a *D*-set. This verifies Condition 4.2.1.b).

(Recall that a subset S of a partially ordered set x is called *order-convex* if $a, c \in S$, $a < c \Rightarrow b \in S$ whenever $a < b < c$.)

c) We have to prove that for $p, q \in I(x, y)$, $p \ll q$ and any forward ray l_p^+ through p, the intersection $l_p^+ \cap \{\beta C_q^- \setminus \{q\}\} \cap I(x, y)$ is nonempty. From part a) above, $I[p, q] \subset I(x, y)$. Since $l_p^+ \cap \{\beta C_q^- \setminus \{q\}\} \subset \beta I[p, q]$, the result follows. The same holds with order reversed.

d) If $p, q \in I(x, y)$, $p \ll q$ and $l \cap \beta C_p^+ \cap I(x, y)$ contains two distinct points, then, since $I(x, y) \subset U$, the result follows from Condition 4.2.1.d).

e) Follows from replacing U by $I(x, y)$ in Condition 4.2.1.e).

f) Follows from replacing U by $I(x, y)$ in Condition 4.2.1.f).

Proof of Part 2: To prove that $I(x, y)$ is l-connected, we have to establish that, for any two distinct points $a_k, b_n \in I(x, y)$, there is an l-polygon from a_k to b_n lying wholly in $I(x, y)$.

Since $y \gg a_k \gg x$ and $y \gg b_n \gg x$, given any forward ray l_x^+ through x, there exist descending polygons $P(a_k, \ldots, a_1, a_0)$ and $P(b_n, \ldots, b_1, b_0)$ that meet $l_x^+ \cap I[x, y]$ at points a_0 and b_0 above x. There are two possibilities: 1) $a_0 \neq b_0$, or 2) $a_0 = b_0$.

If $a_0 \neq b_0$, we may, without loss of generality, assume that $a_0 <^l b_0$. Then $b_1 \gg a_0$, and therefore $\beta C_{b_1}^- \cap l_{a_0, a_1}^+$ is a unique point, say c. Clearly, $y \gg c \gg x$, i.e., $c \in I(x, y)$. If either $c <^l a_1$ or $a_1 <^l c$ on l_{a_0, a_1}, then the l-polygon $P(b_n, \ldots, b_1, c, a_1, \ldots, a_k)$ connects a_k with b_n and lies wholly in $I(x, y)$ (see Fig. 4.5(a)). If $c = a_1$ the same is accomplished by the l-polygon $P(b_n, \ldots, b_1, a_1, \ldots, a_k)$.

Suppose now that $a_0 = b_0$ (see Fig. 4.5(b)). Then $a_2 \gg b_0$, and therefore $l_{b_0, b_1}^+ \cap \beta C_{a_2}^-$ is a unique point, say d, and $\lambda(d, a_2)$. The l-polygon $P(b_n, \ldots, b_1, d, a_2, \ldots, a_k)$ connects a_k with b_n, and lies wholly in $I(x, y)$. ∎

Both parts of Proposition 4.2.17 will be used repeatedly, often without attribution, in the sequel.

f) Remarks on *D*-sets and order intervals

In two-dimensional Minkowski space, U and V, order intervals with sections removed as shown in Fig. 4.6, are *D*-sets. Their intersection $U \cap V$, also shown

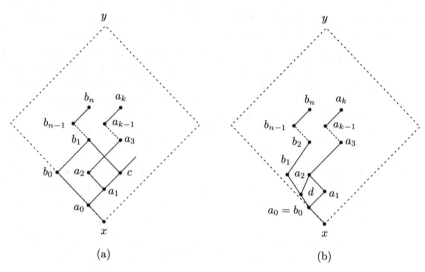

Fig. 4.5. Proving that $I(x,y) \subset U$ is l-connected

in Fig. 4.6, is therefore a D-set, but it is clearly not l-connected. (For emphasis, the two order intervals comprising $U \cap V$ have been slightly shifted with respect to each other.) This example shows that the property of l-connectedness is not stable under intersections. By contrast, open order intervals lying in D-sets are l-connected D-sets.

If a D-set consists of distinct l-connected components, then these components will be pairwise spacelike to each other. Two distinct components cannot contain points that are timelike or lightlike to each other.

g) Incidence theorem for β-boundaries

We now investigate the intersection of two cones such that the vertex of one lies on the β-boundary of the other. We begin with the following lemma:

Lemma 4.2.18 *Let U be a D-set, $x, y \in U$ and $y > x$. Let*

$$x = z_0 < z_1 < \ldots < z_n = y$$

be an increasing l-polygon with $\lambda(z_i, z_{i+1})$, $i = 0, 1 \ldots n - 1$. If the first two light rays l_{z_0,z_1} and l_{z_1,z_2} are distinct, then $y \in \tau C_x^+$. The same statement holds with order reversed.

Proof: Owing to symmetry, it suffices to prove the first statement. By assumption, one has $z_1 \in \beta C_x^+$. If $z_2 \in \beta C_x^+$ then, by the convexity Axiom 4.2.1.d), $l_{z_1,z_2} \subset \beta C_x^+$ and $x \in l_{z_1,z_2}$. Since x and z_1 belong to both light rays, the two rays must be the same, a contradiction. Hence $z_2 \in \tau C_x^+$. ∎

The following incidence theorem follows easily from the above lemma:

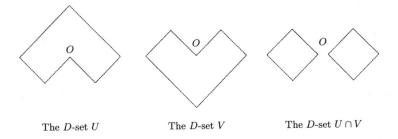

The *D*-set U The *D*-set V The *D*-set $U \cap V$

Fig. 4.6. *D*-sets need not be *l*-connected

Theorem 4.2.19 *Let U be a D-set, $x \in U$ and $p \in \beta C_x^+ \cap U$, $p \neq x$. Then*

1. $\beta C_x^+ \cap \beta C_p^+ \cap U = l_{x,p}^+ \cap \beta C_p^+ \cap U$.
2. $\beta C_x^+ \cap \beta C_p^- \cap U = l[x,p]$.[6]

Proof: Let $a \in \beta C_p^+, p \neq a$. Then $P(x,p,a)$ is an ascending *l*-polygon from x to a (see Fig. 4.7). By Lemma 4.2.18, if $l_{x,p} \neq l_{p,a}$, then $a \in \tau C_x^+$. If $a \notin \tau C_x^+ \cap U$ then necessarily $a \in \beta C_x^+ \cap U$, i.e., $a, p \in \beta C_x^+ \cap U$. Therefore, by the convexity axiom, $x \in l_{p,a}$. However, in a *D*-set two points on a light ray fix the ray uniquely; therefore the ray $l_{x,p}$ is the same as the ray $l_{p,a}$, which proves the first assertion.

To prove the second assertion, note simply that, by the convexity axiom, any point in the intersection $\beta C_x^+ \cap \beta C_p^-$ has to lie on the unique ray that passes through both x and p. ∎

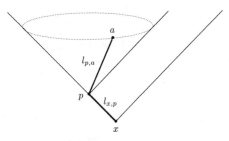

Fig. 4.7. Illustrating Theorem 4.2.19

---

[6] Theorem 4.2.19 is a special case of Proposition 3.2.19 that holds in *D*-sets. To see this, one has to Formulate 3.2.19 entirely in terms of β-boundaries, which is straightforward.

h) Spacelike separation in D-sets

The results established so far in this chapter have been on pairs of points that were lightlike or timelike to each other. We shall now establish two results on spacelike separations in D-sets that will be important in the following.

Theorem 4.2.20 *Let U be a D-set, $x \in U$ and l_x^1, l_x^2 two distinct light rays through x. Choose $a \in \{(l_x^1)^- \cap U\}$, $a \neq x$ and $b \in \{(l_x^2)^+ \cap U\}$, $b \neq x$. Then*

1. $b \gg a$,
2. $I(a,b) \cap C_x = \emptyset$,

i.e., every point of $I(a,b)$ is spacelike to x.

Proof: The fact that $b \gg a$ follows immediately from Lemma 4.2.18. Next, Theorem 4.2.9 ensures that $I(a,b)$ is nonempty. Finally, from Theorem 4.2.19 we have $C_x^+ \cap \tau C_b^- = \emptyset$ and $C_x^- \cap \tau C_a^+ = \emptyset$. Hence $C_x \cap I(a,b) = \emptyset$. ∎

If $r, s \in U$ and $r \ll s$ then, by definition, $C_r^- \cap U \subset C_s^- \cap U$ and the inclusion is such that $\beta C_r^- \cap \beta C_s^- \cap U = \emptyset$; the situation shown in Fig. 4.8(a) cannot arise. If $r \in \beta C_s^-$ and $r \neq s$ then, from Theorem 4.2.19(1), $\beta C_s^- \cap \beta C_r^- \cap U = l_{s,r}^- \cap U$. Therefore, if $\beta C_r^- \cap \beta C_s^- \cap U \neq \emptyset$, and $r \not\ll^{ll} s$, then r and s must be spacelike to each other (Fig. 4.8(b)). Conversely:

Lemma 4.2.21 *Let U be a D-set, $x, r, s \in U$ such that $x \ll r$, $x \ll s$ and r, s are mutually spacelike. Then $\beta C_r^- \cap \beta C_s^- \cap C_x^+$ is nonempty.*

Proof: Let l_x be a light ray through x. From the definition of D-sets, $l_x \cap \beta C_r^-$ is a single point. Call it q_r. Similarly, let $\{q_s\} = l_x \cap \beta C_s^-$. Since q_r and q_s lie on the same light ray, there are three possibilities:

1. $q_r <^{ll} q_s$. Then Lemma 4.2.18 implies $q_r \ll s$, so that the ray $l_{q_r,r}$ intersects βC_s^- at a unique point p, and $p \in \beta C_r^- \cap \beta C_s^- \cap C_x^+$.

2. $q_r = q_s$. Then this point lies, by definition, on $\beta C_r^- \cap \beta C_s^- \cap C_x^+$.

3. $q_s <^{ll} q_r$. This is the same as the case 1) above, with r and s interchanged. ∎

4.3 Timelike Order and D-Subsets

We have already seen (Theorem 4.2.9) that, in a nonempty D-set, every point has timelike predecessors and successors. We shall now establish that analogous results hold for D-subsets. These will lead to the desired separation properties, which will be established in the next section.

Proposition 4.3.1 *Let U, V be D-sets such that $V \subset U$.*

i) *Let $x \in V$ and $z \in U \setminus V$ with $x \ll z$. Then $\exists y \in V$ with $x \ll y \ll z$.*
ii) *The same, with order reversed.*

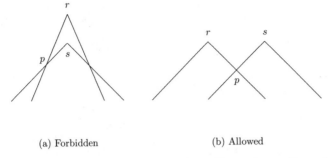

(a) Forbidden   (b) Allowed

Fig. 4.8. Difference between pairs of timelike and spacelike points

Remark: The existence of points y such that $y \in V, y \gg x$ is the content of Lemma 4.2.9. What remains to be proved that there exist such $y \in V$ with $y \ll z$.

Proof: It suffices to prove part i). From Condition 4.2.1.a), $I[x, z] \subset U$. If $I(x, z) \subset V$, then every point $y \in I(x, z)$ satisfies $x \ll y \ll z$. If $I(x, z) \not\subset V$, then for any forward ray l_x^+ from x, there is a backward ray $l_z^{(1)-}$ from z such that $\{q\} = l_x^+ \cap l_z^{(1)-} \in U$ (see Fig. 4.9). Next, let $p \in l_x^+$ such that $x <^l p <^l q$, $x \neq p$, $p \in V$. Finally, let $l_z^{(2)-}$ be a second backward ray from z, and let $\{s\} = l_z^{(2)-} \cap \beta C_p^+$. According to Condition 4.2.1.b), there exist points $y \in l(p, s)$ and $y \neq s$ such that $y \in V$. These points fulfil the requirement $x \ll y \ll z$. ∎

After this preparation, we are able to establish the main result of this section:

Theorem 4.3.2 *Let U, V be D-sets such that $V \subset U$. Let $x \in V$ and $z \in U \backslash V$ with $z \gg x$. Then we can find points u, y, w with $u, y \in V$, $w \in U$ such that*

$$u \ll x \ll y \ll z \ll w.$$

Proof:

a) Apply Theorem 4.2.9 to $x \in V$ to obtain a point $u \in V$ such that $u \ll x$.

b) Apply Theorem 4.2.9 to $z \in U$ to obtain a point $w \in U$ with $w \gg z$.

c) Apply Proposition 4.3.1 to $x \in V$ and $z \in U$, $V \subset U$, to obtain $y \in V$ such that $x \ll y \ll z$.

d) Then $u \ll x \ll y \ll z \ll w$.

∎

We shall call the above theorem the *timelike points theorem*.

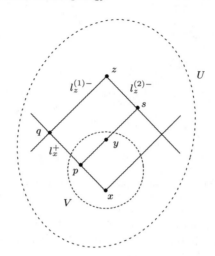

Fig. 4.9. Illustrating Proposition 4.3.1

4.4 Separating Points by D-sets

The aim of this section is to establish Theorem 4.4.2, which states that distinct points in a D-set can be separated by disjoint D-subsets. The proofs utilize a method for transporting "good" properties of one cone to other cones, that is, to spacelike separations.

We begin with the following theorem on separating points by forward and backward cones:

Theorem 4.4.1 Let U be a D-set, $y \in U$ and $b \in U \setminus C_y^-$. Then there exists $a \in U \setminus C_y^-$ such that $b \gg a$.

Proof: There are three possibilities, according to the location of b. They are:

1. $b \gg y$.
2. $b \in \beta C_y^+$.
3. $b \notin C_y^+$.

We establish the existence of the point a case-by-case.

1. By Proposition 4.2.17, if $b \gg y$ then $I(b, y)$ is nonempty. Any point $a \in I(y, b)$ will satisfy $b \gg a$.
2. By the convexity axiom, $b \in \beta C_y^+ \Rightarrow \lambda(y, b)$. Take a point $p \in l_{y,b}$ with $y <^{ll} p <^{ll} b$. Choose a backward ray l_p^- different from $l_{y,b}$, and on it a point $a \in U$ such that $a <^{ll} p$. Then Lemma 4.2.18 implies that $a \ll b$. Since $C_y^- \subset C_p^-$ and $\beta C_y^- \cap \beta C_p^- = l_{y,b} \cap C_y^-$, it follows that $a \notin C_y^-$ (see Fig. 4.10(a)).

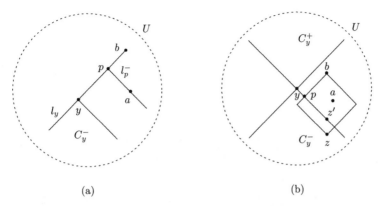

Fig. 4.10. Separation by Forward and Backward Cones

3. Finally, let $b \notin C_y^+$. By Theorem 4.2.9 there exists $z \in U$ such that $z \ll b$.
   (a) If $z \notin C_y^-$ there is nothing to prove (set $a = z$). (b) If $z \in \beta C_y^-$ then,
   by Lemma 4.2.10, there exists $a \in U$ such that $z \ll a \ll b$. Then, by
   Theorem 4.2.20, part 2, $a \notin C_y^-$. c) If $z \in \tau C_y^-$ then, by Lemma 4.2.21,
   there exists $p \in \beta C_y^- \cap \beta C_b^-$. Choose a point $z' \in l_{p,y}$, $z' <^{ll} p$. Then, by
   Lemma 4.2.18, $z' \ll b$ and we are back to the situation (b). ∎

The above separation enables us to construct a separation by D-sets.

Theorem 4.4.2 *If U is a D-set, $x, y \in U$, $x \neq y$ then there exist D-subsets
V and W of U such that $x \in V$, $y \in W$, and $V \cap W = \emptyset$.*

Proof: Since $x \neq y$, the cases $y \in C_x^+$ and $y \in C_x^-$ are mutually exclusive (see
Fig. 4.11). We may therefore assume, without loss of generality, that $y \notin C_x^-$.
By Theorem 4.4.1, there exists a point $p \in U$ such that $p \ll y$ and $p \notin C_x^- \cap U$.
Then $x \notin C_p^+$.

Since $x \notin C_p^+$, we may apply Theorem 4.4.1 with order reversed to x
and C_p^+. This gives us a point $s \in U$ such that $s \gg x$ and $s \notin C_p^+ \cap U$.
Then $C_s^- \cap C_p^+ = \emptyset$. Then Proposition 4.3.1 tells us that there exist points
$r, q \in U$ such that $p \ll y \ll q$, $r \ll x \ll s$. Then $x \in I(r, s)$, $y \in I(p, q)$,
$I(p, q) \cap I(r, s) = \emptyset$, and $I(p, q)$, $I(r, s)$ are D-subsets of U. ∎

4.5 Local Structure and Topology

In the preceding sections we have defined D-sets and have established their
fundamental properties. They lead us very naturally to the local structure
axiom and the topology.

Axiom 4.5.1 (The local structure axiom) *The ordered space M satisfies
the following axiom: For each $x \in M$ there exists a D-set U_x such that $x \in
U_x \subset M$.* □

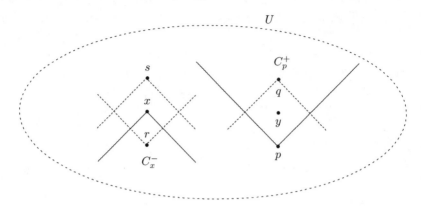

Fig. 4.11. Separation by D-Sets

Definition 4.5.2 The **order topology** on M is defined to be that topology which has the family of D-subsets as a base. □

Remarks 4.5.3

1. It follows from Proposition 4.2.16 (the intersection of two D-sets is a D-set) that the family of D-subsets of M is indeed a base for a topology on M.
2. Theorem 4.4.2 now states that the order topology is Hausdorff.
3. A D-set is defined by its properties, and the order topology is defined using *all* subsets of M that fulfil the defining conditions of D-sets. At first sight, it might appear that one could define a finer order topology by throwing away some "large" D-sets. However, this does not work, because the sets that have been thrown away reappear as unions of D-sets properly contained in them, and the set so defined will obviously satisfy all conditions required of D-sets.

Theorem 4.5.4 *In every D-set, $\beta C_x^+ = \partial C_x^+$ (the boundary of C_x^+), and $\tau C_x^+ = \operatorname{int} C_x^+$ (the interior of C_x^+), and the same for reversed order.*

Proof: Only one of the assertions needs to be proven (as it would automatically imply the other). If $y \in \tau C_x^+$, then there exist points $a, b \in C_x^+$ such that $x \ll a \ll y \ll b$, i.e., $y \in I(a, b)$. Hence $\tau C_x^+ = \operatorname{int} C_x^+$. ∎

The two lemmas that follow are intuitively clear, and often useful:

Lemma 4.5.5 *The family of open order intervals*

$$\{I(x, y) \mid x, y \in U, \ x \ll y, \ U \text{ is a } D\text{-set}\}$$

is a base for the order topology.

Proof: Let I_1, I_2 be two such order intervals with $I_1 \cap I_2 \neq \emptyset$. By Proposition 4.2.16, $I_1 \cap I_2$ is a D-set. Let $p \in I_1 \cap I_2$. By Theorem 4.2.9, there exists a point $q \in I_1 \cap I_2$ such that $p \ll q$. Then $I(p, q) \in I_1 \cap I_2$. ∎

Lemma 4.5.6 *An open subset U of an ordered space M is a D-set iff it fulfils the following conditions:*

i) U is order-convex.

ii) If $x, z \in U$ with $x \ll z$ and $y \in I(x, z)$ then every light ray l_y intersects $\beta I[x, z]$ in two points u, v with $u <^l y$ and $y <^l v$.

iii) If $a, b \in U$ and $\lambda(a, b)$ then the ray $l_{a,b}$ is unique.

Proof: If U is a D-set, then the above conditions are satisfied by definition. Conversely, we have to verify that the above conditions imply the properties a)–f) of Def. 4.2.1.

a) Let $x, y \in U$. Then, by the definition of order-convexity (see Def. 4.2.1.a), $I[x, y] \subset U$.

b) Let $l \cap U \neq \emptyset$. Suppose that $l \cap U$ has a maximum, say b. As U is open, b is an interior point of U. Therefore there exists a D-set V_b such that $b \in V_b \subset U$, and therefore b is a maximum for $l \cap V_b$, which contradicts the defining Condition 4.2.1.b) of D-sets. By the same argument, $l \cap U$ cannot have a minimum.

c) This follows immediately from condition ii).

d) Let $a, b \in \beta C_x^+$ with $a < b$. Then there exists an ascending l-polygon $P(c_0, c_1, \ldots c_n)$ such that adjacent c_i lie in a D-set, and $x = c_0$, $b = c_n$ and $a = c_k$ for some k, with $0 < k < n$. We shall denote $l_{c_i, c_{i+1}}$ by $l_{i,i+1}$. If $l_{0,1}$ is different from $l_{1,2}$ and V is a D-set containing c_1, then there exist points $p \in l_{0,1} \cap V$ and $q \in l_{1,2} \cap V$. Since $c_1 \in \beta C_p^+$, it follows from Lemma 4.2.18 that $q \in \tau C_p^+ \subset \tau C_x^+$. From $q <^l c_2$ it then follows that $c_2 \in \tau C_p^+ \subset \tau C_x^+$, which contradicts our assumption. Therefore we must have $l_{1,2} = l_{0,1}$ and $c_2 \in \beta C_x^+$. Let now $l_{0,1} = \cdots = l_{i,i+1}$ and $c_1, \ldots, c_{i+1} \in \beta C_x^+$. If $l_{i+1,i+2} \neq l_{i,i+1}$ then $c_{i+2} \in \tau C_i^+$ and therefore $b \in \tau C_x^+$. It follows that $l_{i,i+1} = l_{i+1,i+2}$. Therefore, by induction, all $c_i \in \beta C_x^+$. Since the choice of the c_i is arbitrary except for the condition that adjacent c_i lie in a D-set, we conclude that all points of $l_{a,b}$ which lie between a and b belong to βC_x^+.

e) This is the same as condition iii).

f) This holds because every point of M lies in a D-set. ∎

Remark 4.5.7 The order topology introduced above clearly coincides with the standard topology on $\mathbb{R}^4$ in Minkowski space, and is therefore strictly coarser than the fine topology on Minkowski space introduced by Zeeman [133]. By itself, the order topology does not imply a "causal" or a linear structure on M.

4.6 Regularity and Complete Regularity

In this section we shall establish that ordered spaces have the functional separation property of Urysohn's lemma. We begin with a remark on terminology.

Remark 4.6.1 There is no agreement in the literature, even at the textbook level, on the use of the terms *regular, completely regular* and *Tychonoff*. Our use of these terms will conform with that of Willard [130] and Kelley [57]. That is, (i) a regular space will be one in which a point and a closed set disjoint from it can be separated by disjoint open sets; (ii) a completely regular space will be one in which a point and a closed set disjoint from it can be separated by a continuous real-valued function; (iii) a Tychonoff space will be one which is completely regular, and in which one-point sets are closed.[7]

Theorem 4.6.2 (Complete regularity) *Let M be an ordered space, $b \in M$ and A a closed subset of M such that $b \notin A$. Then there exists a continuous function $f : M \to [0,1]$ such that $f(x) = 0$ and $f(A) = 1$.*

Proof: The proof is exactly the same as that of Urysohn's lemma, except that, instead of normality, one uses Lemma 4.2.10 to obtain a family of open sets with the required nesting property.

In the following, I will denote an open D-interval $I(.,.)$, and $\bar{I}$ its closure $I[.,.]$. Next, let $\mathbb{P} = \mathbb{Q} \cap [0,1]$, and let an enumeration $p_0, p_1, \ldots p_n, \ldots$ of $\mathbb{P}$ be given such that $p_0 = 0$ and $p_1 = 1$. Set $\mathbb{P}_n = \{p_0, \ldots p_n\}$.

Since $b \in M \setminus A$ and $M \setminus A$ is open, there exists an open D-interval I_1 such that $b \in I_1 \subset M \setminus A$. By Lemma 4.2.10, there exist points $x, y \in I_1$ such that $x \ll b \ll y$, i.e., $b \in I(x,y) \subset I[x,y] \subset I_1$. Set $I_0 = I(x,y)$.

Suppose that, for $r \in \mathbb{P}_n$, D-intervals I_r that have the property

$$r < s \Rightarrow \bar{I}_r \subset I_s \quad \text{for} \quad r, s \in \mathbb{P}_n \tag{4.1}$$

have been defined. We shall define a D-interval I_r for $r = p_{n+1}$ such that the property (4.1) holds for $r, s \in \mathbb{P}_{n+1}$.

Since P_n contains 0 and 1 and $0 < p_{n+1} < 1$, the number p_{n+1} partitions $\mathbb{P}_n$ into two disjoint subsets L and R such that $r \in L \Rightarrow r < p_{n+1}$, and $r \in R \Rightarrow r > p_{n+1}$. Since $\mathbb{P}_n$ is finite, L has a largest member w and R a smallest member u, and

$$w < p_{n+1} < u .$$

---

[7] Some authors (such as Munkres, [77]) include T_1-separation as an integral part of the definition of regular and completely regular spaces. Yet other conventions are used in the reference handbook [101].

The D-intervals I_u and I_w are already defined. Let $u_1, u_2, w_1, w_2 \in I_1$ such that $I_u = I(u_1, u_2)$ and $I_w = I(w_1, w_2)$. Then

$$u_1 \ll w_1 \ll w_2 \ll u_2 .$$

By Lemma 4.2.10, there exist $v_1, v_2 \in I_1$ such that

$$u_1 \ll v_1 \ll w_1 \ll w_2 \ll v_2 \ll u_2 .$$

Then $\bar{I}_v \subset I_u$ and $\bar{I}_w \subset I_v$.

The above procedure defines, recursively, a set of D-intervals I_r, indexed by $r \in \mathbb{P}$, that have the properties

1. $b \in I_0$;
2. $r < s \to \bar{I}_r \subset I_s \ \forall \ r, s \in \mathbb{P}$;
3. $I_1 \cap A = \emptyset$.

Define now the function

$$f(x) = \begin{cases} 1, & \text{if } x \notin \text{any } U_r; \\ \inf \{ r | x \in U_r \}, & \text{otherwise.} \end{cases} \tag{4.2}$$

Clearly, $f(b) = 0$ and $f(A) = 1$. The function $f(x)$ is an Urysohn function, and the proof of its continuity is identical with that in the proof of Urysohn's lemma. See [130] or [77]. ■

Since one-point sets are closed in M, it follows immediately that

Corollary 4.6.3 *The ordered space M is a Tychonoff space.*[8] ■

These results will be exploited in Chap. 6.

4.7 Order Equivalence

Let M, M' be spaces fulfilling the order, identification and cone Axioms 3.1.2, 3.1.12 and 3.2.8. A map $f : M \to M'$ will be called an *order equivalence* iff

1. f is bijective.
2. Under f, the image of a light ray in M is a light ray in M'.
3. $a, b \in M, a <^{ll} b$ implies that $f(a) <^{ll} f(b)$ in M'.
4. If l^1, l^2 are light rays in M, then $f(l^1 \cap l^2) = f(l^1) \cap f(l^2)$ whenever l^1 and l^2 intersect in M.

---

[8] It is a standard result in point-set topology that every locally compact Hausdorff space is a Tychonoff space. See, for example, [130]. However, an ordered space M need not be locally compact.

5. If $D \subset M$ is a D-set, then so is $f(D) = D' \subset M'$.

Order equivalence is clearly an equivalence. It would be tempting to regard two order-equivalent spaces as essentially identical, but, as we shall see in Sect. 8.4, the order structures of two order-equivalent spaces may be differentially inequivalent. However, different order structures may give rise to the same topology, and metrically distinct spaces may be order equivalent:

Examples 4.7.1

1. The de Sitter space in two dimensions is a hyperboloid of one sheet, and is homeomorphic with the cylinder $S_1 \times \mathbb{R}$. Light rays in the de Sitter space are the (two) families of generators of the hyperboloid. These are straight lines, and a generator from one family intersects a generator from the other family once, and only once.

   We give an order structure to $S_1 \times \mathbb{R}$ as follows: Imagine the cylinder to be placed vertically, and define the light rays to be curves that make an angle of $\pi/4$ with the horizontal circular sections of the cylinder. There are two light rays through any point of $S_1 \times \mathbb{R}$, and these two rays intersect infinitely many times. The ordered cylinder is not order-equivalent to the de Sitter space in two dimensions.

2. The punctured plane is homeomorphic with the plane from which a closed circular disk with centre at the origin has been excised. Now take two copies of the two-dimensional Minkowski space, excise the origin from the first and a closed circular disk around the origin from the second. Excising a point cuts exactly *two* light rays into two light rays each; excising a disk does it to *infinitely many* light rays. This the two spaces cannot be order-equivalent.

3. Two-dimensional Minkowski space $\mathbb{M}^2$ and the two-dimensional wedge W defined by $|x_0| < x_1$ are homeomorphic and order equivalent. To see this, furnish Minkowski space with the light-cone coordinates $x^\pm$ defined by $x^\pm = x_0 \pm x_1$, and set $w^\pm = \exp x^\pm$. This map sends light rays to light rays and D-sets to D-sets.

   In physics the spaces $\mathbb{M}^2$ and W are regarded as different because they are not isometric.

5

Homogeneity Properties

Homogeneity properties of D-sets give rise to strong homogeneity properties on i) light rays, ii) intersections of boundaries of backward and forward cones, and iii) D-intervals. We shall study the first two of these properties systematically in this chapter. The main strategy is to study the property in question in a D-set, and then try to extend it beyond the given D-set. The attempt to do so reveals a number of new features, and we shall discuss them in detail.

These methods are not powerful enough to establish the corresponding homogeneity property of D-intervals themselves, and an additional tool is needed. Although the result is valid for ordered spaces as defined earlier, we have not been able to develop this tool (the existence of timelike curves) for spaces that are not complete. The completion of ordered spaces will be studied in Chap. 6, and its consequences elaborated in Chap. 8.

5.1 Light Rays and D-sets

We begin with the study of homogeneity properties of light ray segments in D-sets by means of natural maps mediated by (other) light rays. We first establish a useful auxiliary result on the topology of a light ray.

Proposition 5.1.1 *The subspace topology on l induced by the order topology of M is the standard order topology on l.*

Proof: By definition, the subspace topology on l has, for a base, the intersections of l with the open order intervals $I(x, y)$ that are subsets of D-sets. Consider a *closed* order interval $I[x, y]$ that intersects l. The intersection either lies wholly on the boundary of $I[x, y]$, or meets it at exactly two points. The intersection of l with the open order interval $I(x, y)$ is therefore an open light-ray segment. The family of these open segments is a base for the standard order topology on l. ∎

H.-J. Borchers and R.N. Sen: *Mathematical Implications of Einstein–Weyl Causality*,
Lect. Notes Phys. **709**, 51–65 (2006)
DOI 10.1007/3-540-37681-X_5

5.1.1 Homeomorphism of Light Ray Segments

A number of useful results follow rather easily from Proposition 5.1.1. We
begin with a definition, and a result on natural maps of light-ray segments in
D-sets (Fig. 5.1).

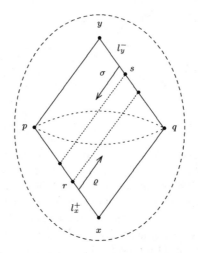

Fig. 5.1. The Standard Maps ϱ and σ

Definition 5.1.2 Let U be a D-set, $x, y \in U$, $y \gg x$ and $l_x \ni x$. Set

$$l_x \cap \beta C_y^- = \{p\} \ .$$

Let $l_y \ni y$, $l_y \neq l_{p,y}$, and set

$$l_y \cap \beta C_x^+ = \{q\} \ .$$

Let now

$$r \in l_x, \ x <^l r <^l p \ ;$$
$$s \in l_y, \ q <^l s <^l y \ .$$

Define the maps

$$\varrho : l_x[x, p] \to l_y$$

and

$$\sigma : l_y[q, y] \to l_x$$

by

$$\varrho(r) = l_y \cap \beta C_r^+ \tag{5.1}$$

and

$$\sigma(s) = l_x \cap \beta C_s^- \tag{5.2}$$

respectively. These maps are well-defined, since the right-hand sides of (5.1) and (5.2) are unique points. $\qquad\square$

Proposition 5.1.3 *The maps ϱ and σ defined by (5.1) and (5.2) are a) order-preserving, and b) bijective; furthermore, c) $\sigma = \varrho^{-1}$ and $\varrho = \sigma^{-1}$.*

Proof:

a) Consider ϱ. The relation $r <^l r'$ implies $C_{r'}^+ \subset C_r^+$, and therefore

$$l_y[\varrho(r), y] \supset l_y[\varrho(r'), y] ,$$

which implies $\varrho(r) <^l \varrho(r')$. Similarly for σ.

b) Let $r, r' \in l_x[x, p]$, $r <^{ll} r'$. Then, by a) above, $\varrho(r) <^l \varrho(r')$. Suppose that $\varrho(r) = \varrho(r') = t$ (Fig. 5.2). The rays $l_{r,t}$ and $l_{r',t}$ both lie on βC_t^-, i.e., $r, r' \in \beta C_t^-$. But as $\lambda(r, r')$, it follows from the convexity axiom that $l_{r,r'}$ lies on βC_t^- and passes through t, which contradicts the construction. Therefore $\varrho(r) \neq \varrho(r')$. Similarly $s \neq s' \Rightarrow \sigma(s) \neq \sigma(s')$.

c) From Lemma 4.2.12, it follows that every $r \in l_x[x, p]$ has a unique σ-image $s \in l_y[q, y]$, and vice-versa. Since, in a D-set, two points can be joined by at most one light ray (Def. 4.2.1.e), it follows that $\varrho(r) = s \Leftrightarrow \sigma(s) = r$. $\qquad\blacksquare$

These bijective maps are actually homeomorphisms:

Proposition 5.1.4 *Let U be a D-set, $x, y \in U$, $x \ll y$ and $l_x \ni x$. Set*

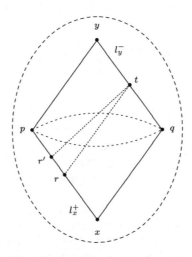

Fig. 5.2. Proving that ϱ and σ are homeomorphisms

$$l_x \cap \beta C_y^- = \{p\} .$$

Let $l_y \ni y$, $l_y \neq l_{p,y}$ and set

$$l_y \cap \beta C_x^+ = \{q\} .$$

Then the segments $l_x[x,p]$ and $l_y[q,y]$ are homeomorphic.

We shall denote this as follows: $l_x[x,p] \stackrel{\text{hom}}{=} l_y[q,y]$.

Proof: Follows from the basic fact that a bijection of a totally ordered set which is order-preserving in both directions is a homeomorphism in the topology defined by the total order. ∎

We shall call the maps ϱ and σ defined above *standard maps*.

From now on, we shall assume a little bit more; we want to exclude the "two-dimensional" case, for which the results that follow will not hold:

Assumption 5.1.5 *There are at least three distinct light rays through each point of U.*

From here on, Assumption 5.1.5 will be deemed to hold unless the contrary is explicitly stated. When it holds, there are enough natural maps in a D-set to prove that any two light-ray segments (in a D-set) are homeomorphic to each other. We break up the proof into three propositions.

Proposition 5.1.6 *Let U be a D-set, $x,y \in U$ and $x \ll y$. Let l_x^1, l_x^2, l_x^3 be three distinct light rays through x, which are intersected, respectively, by the rays l_y^1, l_y^2 and l_y^3 through y. Let the points of intersection be p_1, p_2 and p_3 respectively (See Fig. 5.3). Then p_1, p_2 and p_3 are distinct, and the six closed segments $l_x^1[x,p_1], l_x^2[x,p_2], l_x^3[x,p_3], l_y^1[p_1,y], l_y^2[p_2,y], l_y^3[p_3,y]$ are homeomorphic to each other.*

Proof: The fact that p_1, p_2 and p_3 are distinct is trivial. Applying Proposition 5.1.4 repeatedly, we find that $l_y^2[p_2,y] \stackrel{\text{hom}}{=} l_x^1[x,p_1] \stackrel{\text{hom}}{=} l_y^3[p_3,y] \stackrel{\text{hom}}{=} l_x^2[x,p_2] \stackrel{\text{hom}}{=} l_y^1[p_1,y] \stackrel{\text{hom}}{=} l_x^3[x,p_3]$. ∎

If we call the segments $l_x[x,p]$ and $l_y[p,y]$ *boundary segments* of the order interval $I[x,y] \subset U$, the above result can be paraphrased as follows: *In a D-set, any two boundary segments of an order interval are homeomorphic to each other.* The same is true for the open segments obtained by deleting the end-points.

Observe that the argument *fails* if there are only two light rays through each point of M; there are not enough natural maps. However, in this case our main goal, the differential structure – when it exists – is attained without much effort, as we shall see later. Anticipating future results, this case will be called the *two-dimensional case*.

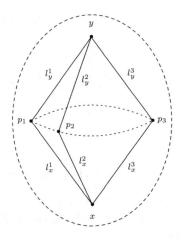

Fig. 5.3. Proving Proposition 5.1.6

Our next result establishes that, in an l-connected D-set, a closed light ray segment is homeomorphic to any closed subsegment of it.

Proposition 5.1.7 *Let U be a D-set, $x \in U$ and l_x a light ray through x. Let $p', p \in l_x \cap U$ such that $x <^{ll} p' <^{ll} p$. Then*

$$l_x[x, p'] \stackrel{\mathrm{hom}}{=} l_x[x, p] \stackrel{\mathrm{hom}}{=} l_x[p', p] .$$

Proof: Let $y \in \beta C_p^+ \cap U$, $y \notin l_{x,p}$. Then $y \gg x$ and $I[x, y] \subset U$. Let $p, q \in \beta C_x^+ \cap \beta C_y^-$, $p \neq q$. Finally, let $\{q'\} = \beta C_{p'}^+ \cap l_{q,y}$, and write (see Fig. 5.4)

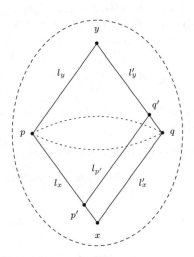

Fig. 5.4. Homeomorphism of closed subsegments

$l'_x = l_{x,q}, l_y = l_{p,y}, l_{y'} = l_{q,y}$ and $l_{p'} = l_{p',q'}$. Apply Proposition 5.1.6 repeatedly, as follows:

1. to $I[x, y]$ to obtain $l_x[x, p] \overset{\text{hom}}{=} l_y[p, y]$;

2. to $I[p', y]$ to obtain $l_x[p', p] \overset{\text{hom}}{=} l_y[p, y] \overset{\text{hom}}{=} l_{p'}[p', q']$;

3. to $I[x, q']$ to obtain $l_x[x, p'] \overset{\text{hom}}{=} l_{p'}[p', q']$.

The required result follows from the above.   ■

Next, we extend the above result to l-polygons in a D-set:

Proposition 5.1.8 *Let U be a D-set, and let $P(x_0, x_1, \ldots, x_n)$ be an l-polygon lying wholly in U. Then the light-ray segments*

$$l[x_i, x_{i+1}], \quad i = 0, 1, \ldots, n-1$$

are homeomorphic to each other.

Proof: By induction. Suppose that the segments $l[x_j, x_{j+1}]$ are homeomorphic to each other for $j = 0, 1, \ldots, k$, $k < n-1$. We shall prove that the segment $l[x_{k+1}, x_{k+2}]$ is homeomorphic to the segment $l[x_k, x_{k+1}]$.

The three points x_k, x_{k+1}, x_{k+2} must satisfy one of the following conditions:

1. $x_k <^{ll} x_{k+1} <^{ll} x_{k+2}$ (or the same, with order reversed); or

2. $x_k, x_{k+2} \in \beta C^-_{x_{k+1}}$ (or the same, with order reversed).

If 1) holds, then $I[x_k, x_{k+2}] \subset U$. Choose a point $p \in \beta C^+_{x_k} \cap \beta C^-_{x_{k+2}}$. Proposition 5.1.6 applied to the closed 4-gon $P(x_k, x_{k+1}, x_{k+2}, p, x_k)$, which lies on $\beta I[x_k, x_{k+2}]$, gives the desired result (Fig. 5.5(a)).

If 2) holds, then there exist points $z \in l_{x_k, x_{k+1}} \cap U$ such that $x_{k+1} <^{ll} z$ (Fig. 5.5(b)). Then $z \gg x_{k+2}$ and $I[x_{k+2}, z] \subset U$. From Proposition 5.1.6, $l[x_{k+1}, z] \overset{\text{hom}}{=} l[x_{k+2}, x_{k+1}]$, and from Proposition 5.1.7, $l[x_{k+1}, z] \overset{\text{hom}}{=} l[x_k, x_{k+1}]$, from which it follows that $l[x_k, x_{k+1}] \overset{\text{hom}}{=} l[x_{k+2}, x_{k+1}]$.

This establishes the inductive step. The proof of the initial step ($j = 0, 1$) is identical with the above.   ■

Our final result in this direction is the following:

Theorem 5.1.9 *Any two closed light-ray segments in an l-connected D-set U are homeomorphic to each other.*

Proof: Let $l[x_0, x_1] \subset U$ and $l[y_0, y_1] \subset U$ be two light ray segments. If they intersect, choose a point p on $\beta C^-_{y_0} \cap U$ such that $p \neq y_0$ and $p \notin l[x_0, x_1]$ (Fig. 5.6). Then $I[p, y_1] \subset U$ and therefore, by Proposition 5.1.6, $l[p, y_0] \overset{\text{hom}}{=}$

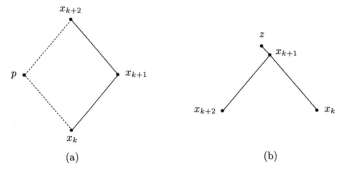

Fig. 5.5. Proving Proposition 5.1.8

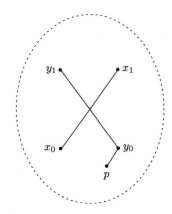

Fig. 5.6. A step in the proof of Theorem 5.1.9

$l[y_0, y_1]$. It remains to prove that two non-intersecting segments $l[x_0, x_1]$ and $l[y_0, y_1]$ are homeomorphic.

Since $x_1, y_0 \in U$, there exists an l-polygon $P(x_1, a_1, \ldots, a_n, y_0)$ lying wholly in U. We may, without loss of generality, assume that the segments $l[x_0, x_1]$ and $l[y_0, y_1]$ do not form part of $P(x_1, a_1, \ldots, a_n, y_0)$. Then $P(x_0, x_1, a_1, \ldots a_n, y_0, y_1)$ is an l-polygon lying wholly in U. The result now follows from Proposition 5.1.8. ∎

Remark 5.1.10 Since M does not have boundaries (every point of M lies in a D-set), the result proven above for closed light ray segments holds also for open segments. The proof is straightforward, and we omit the details.

Clearly, Theorem 5.1.9 can be extended outside the D-set U if there is a D-set V which overlaps U and covers a part of l that is not inside U. Our next step will therefore be to study the conditions under which such extensions are possible. It will be seen that care has to be exercised in dealing with spaces that are not complete.[1] We begin by recapitulating a few definitions and results from point-set topology in the context of ordered spaces, and providing a few examples.

5.2 Topological Preliminaries

We first define the notion of an *overlapping cover* of a light ray by D-sets.

Definition 5.2.1 Let M be an ordered space,[2] l a light ray in M, and $\mathcal{U} = \{U_\alpha | \alpha \in A\}$ a covering of l by D-sets. Here A is some indexing set. We say that $\mathcal{U}$ is an *overlapping cover* of l if, for any pair of points $x, y \in l, x <^{ll} y$, there exist points $x = x_0, x_1, \ldots x_n = y$ on l, $x_i <^{ll} x_{i+1}, i = 0, 1, \ldots n-1$ and D-sets $U_k \in \mathcal{U}$ such that $l[x_k, x_{k+1}] \subset U_k$. □

Next, we recall the following definitions:

1. Let X be a topological space. A *separation* (or *disconnection*) of X is a pair of nonempty open sets U and V such that $U \cap V = \emptyset$ and $U \cup V = X$. If there exists no separation of X, then X is said to be connected.
2. A topological space is called *totally disconnected* if the only connected subsets are the one-point subsets.
3. The topology of X is called *discrete* if every subset of X is open.

Let us now consider the two examples that follow in some detail:

Example 5.2.2 Let $M = \mathbb{Q}^2$ and let the light rays be the lines $x - y = \text{const}$ and $x + y = \text{const}$, $x, y \in \mathbb{Q}$. Define the order topology $\mathcal{T}$ using as base the family of order intervals[3]

$$\mathcal{B} = \{I(p, q); p, q \in M, p \ll q\} . \tag{5.3}$$

This topology endows M with the following properties:

1. M is totally disconnected. For let $a, b \in M, a \neq b$. Let $(a_x, a_y), (b_x, b_y)$ be the coordinates of a and b respectively. Suppose that $a_x \neq b_x$. Choose an irrational number ξ lying between a_x and b_x, and consider the sets

$$M_A = \bigcup_{\substack{p \ll q \\ p_x > \xi}} I(p, q) ,$$

$$M_B = \bigcup_{\substack{p \ll q \\ q_x < \xi}} I(p, q) . \tag{5.4}$$

---

[1] The word *complete* is used in this chapter purely as a guide to intuition, as it would be used in a metric or uniform space. We shall discuss the uniformizability and completion of ordered spaces in Chap. 6.

[2] Recall that M is assumed to be l-connected.

[3] In n dimensions, $n > 2$, one has to replace $\mathbb{Q}$ by a real algebraic extension $\mathbb{F}$ that is closed under the taking of square roots.

Then M_A and M_B form a separation of M. Next, let $a_x = b_x$. Then $a_y \neq b_y$. Choose an irrational η lying between a_y and b_y, and define

$$M_L = \{ \bigcup_{\text{all } I(p,q)} I(p,q) | a \in I(p,q) \Rightarrow a_x < \eta \} \, ,$$

$$\text{(5.5)}$$

$$M_R = \{ \bigcup_{\text{all } I(p,q)} I(p,q) | a \in I(p,q) \Rightarrow a_x > \eta \} \, .$$

Then M_L and M_R form a separation of M. That is, no two distinct points of M lie in a connected subset, i.e., M is totally disconnected.

2. M is not discrete. For, the open sets in M are arbitrary unions of finite intersections of the basis elements defined by (5.3). By definition, if $x \in B_1$, $x \in B_2$, then there exists a basis element, i.e., a nonempty order interval B_3 such that $x \in B_3 \subset B_1 \cap B_2$. That is, if the intersection of a finite number of basis elements is nonempty, then it contains a nonempty order interval. Therefore one-point sets cannot be open in M.

3. Every light ray in M has an overlapping cover,[4] which can be constructed as follows. Let l be a light ray, and l^1, l^2 two light rays that are parallel to l, and are different from l and from each other. Then l lies in the strip bounded by the parallel lines l^1 and l^2 in $\mathbb{Q}^2$. We may assume that l^2 lies above l^1, without loss of generality. Let now $a \in l^1$ and $b \in l^2$. The family of open order intervals $\{I(a,b) | a \in l^1, b \in l^2\}$ is an overlapping cover of l. (Recall that $I(a,b) = \emptyset$ if $a \not\ll b, a \neq b$.)

4. *Not every cover of l has an overlapping subcover.* Examples are covers chosen from the separations $M = M_A \cup M_B$ and $M = M_L \cup M_R$ defined by (5.4) and (5.5) respectively.[5]

Example 5.2.3 Let W (the double wedge) be the complement of the dotted cone bounded by the straight lines $y = 2x - \pi$ and $y = -2x + 3\pi$ in the Minkowski plane $M = \mathbb{R}^2$, as shown in Fig. 5.7. These two lines intersect at the point $p = (\pi, \pi)$. Any D-set in M that contains the point p contains points of W, as well as points of $M \setminus W$. Let $X = W \cap \mathbb{Q}^2$. Define light rays in X to be the restrictions of light rays in M to X. Then X becomes an ordered space.

X consists of two disjoint pieces L and R that lie to the left and right, respectively, of W. Although disjoint, the pieces L and R are connected by the two light rays l^1 and l^2, shown by dashed lines in the figure. l^1 is the line $y = x$, and l^2 the line $y = -x + 2\pi$. That is, X is l-connected, but the rays l^1, l^2 do not have overlapping covers.

---

[4] It is worth emphasizing that the existence of overlapping covers (of light rays) is *not* a universal property. Example 5.2.3 that follows provides a counterexample.

[5] The significance of the last observation lies in the fact that if every cover of l had an overlapping subcover, then l would have been homeomorphic to $\mathbb{R}$; this is a standard theorem of point-set topology.

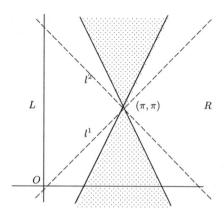

Fig. 5.7. Light rays that have no overlapping covers

In X, owing to the excision of the point p, the line l^1 splits into two distinct light rays, each of which has an overlapping cover, and the same is true of the line l^2.

5.3 Segments of Light Rays

In this section we shall show how, in the presence of overlapping covers, Proposition 5.1.9 may be extended to the entire space. We shall also prove analogous results for spacelike hyperspheres defined by (5.6) below. Results for order intervals that lie in D-sets require further preparation, and will be dealt with in the following section.

We shall describe these results by the term *homogeneity properties* of the ordered space M.

It is worth noting that the results of this section and the next depend only on the existence of overlapping covers for light rays; it does not matter if, like $\mathbb{Q}^2$ in the topology $\mathcal{T}$, the space is totally disconnected.

Our first result is the following:

Proposition 5.3.1 *Let l be a light ray in the ordered space M. If l has an overlapping cover $\mathcal{U}$, and $U_x, U_y \in \mathcal{U}$, then any closed segment of l in U_x is homeomorphic to any closed segment of l in U_y.*

Proof: If $U_x \cap U_y \neq \emptyset$, then there exists a closed segment $l[a, b]$ of l which is contained in both U_x and U_y. By Theorem 5.1.9, $l[a, b]$ is homeomorphic to every closed segment of l contained in U_x, as well as to every closed segment contained in U_y.

If $x \in l \cap U_x$, $y \in l \cap U_y$, $x \neq y$ and $U_x \cap U_y = \emptyset$ then we may, without loss of generality, assume that $x <^l y$. Since l has an overlapping cover, there exist points $x = x_0, x_1, \ldots x_{n-1}, x_n = y$ on l and D-sets $D_x = D_0, D_1, \ldots D_n = D_y$ such that $l[x_k, x_{k+1}] \subset U_k$. Since $D_k \cap D_{k+1} \neq \emptyset$, the result follows by induction. $\blacksquare$

It is worth recalling that the proof of Proposition 5.1.9 is based, ultimately, upon the topological theorem that a bijective order-preserving map between two totally ordered sets is a homeomorphism (in their order topologies).

The general result follows quickly from the above:

Theorem 5.3.2 (First homogeneity property: $l[a, b]$) *Let M be an ordered space in which every light ray has an overlapping cover. Let $U_1, U_2 \subset M$ be D-sets and $l^1[a, b] \subset U_1$, $l^2[p, q] \subset U_2$ be nonempty closed light ray segments. Then*

$$l^1[a, b] \stackrel{\mathrm{hom}}{=} l^2[p, q] \; .$$

Proof: Since M is l-connected, there exists an l-polygon connecting b with p. By assumption, every light ray has an overlapping cover, and therefore the l-polygon $P(a, b, \ldots, p, q)$ has an overlapping cover. The result now follows from Proposition 5.3.1. $\blacksquare$

As remarked earlier, in an ordered space in which light rays are locally homeomorphic with $\mathbb{R}$, every light ray has an overlapping cover. This condition may fail only when light rays do not have the local structure of the linear continuum, as in Example 5.2.3. When it fails, there is no assurance that the first homogeneity Property 5.3.2 will hold. We wish to exclude such situations, and do so by the following explicit assumption:

Assumption 5.3.3 (Overlapping cover assumption)

The ordered space M will be assumed to be such that every light ray has an overlapping cover.

5.4 Spacelike Hyperspheres

We begin with a remark on terminology and notations. In a Minkowski space, if $p \ll q$, then the intersections of the boundaries

$$S(p, q) = \beta C_p^+ \cap \beta C_q^- . \tag{5.6}$$

are spacelike hyperspheres. By abuse of language, we shall use the same term to denote the intersections (5.6) of boundaries of forward and backward cones *inside D-sets* in an arbitrary ordered space M.

We shall establish that spacelike hyperspheres defined by (5.6) are homeomorphic to each other. As with homeomorphisms of light ray segments, the proof will be broken up into several steps. We begin with the following technical lemma:

Lemma 5.4.1 *Let $\{U_\alpha\}, \alpha \in A$ be an indexed family of D-sets. Let $Z = \cap_{\{\alpha \in A\}} U_\alpha$. Then int Z is a D-set.*

Proof: As an intersection of D-sets, Z may fail to be a D-set only because the infinite intersection introduces boundary points. These are eliminated by taking the interior. Recall that the empty set is trivially a D-set.   ∎

Proposition 5.4.2 *Let U be a D-set, $a, b, c \in U, a \ll b \ll c$. Then*

$$S(a,b) \overset{\text{hom}}{=} S(a,c) \overset{\text{hom}}{=} S(b,c) .$$

Proof: By part c) of Definition 4.2.1, the map[6] $\pi : S(a,b) \to S(a,c)$ mediated by light rays through a is bijective (Fig. 5.8). We have to prove that π and π^{-1} are continuous.

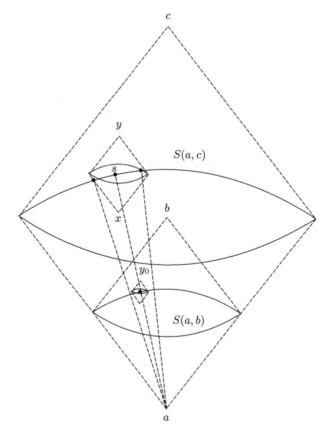

Fig. 5.8. Proving that $S(a,b)$ and $S(a,c)$ are homeomorphic

---

[6] Observe that this map is a stereographic projection.

We shall prove continuity in its local form: the map $\pi : S(a,b) \to S(a,c)$ is continuous at the point $s_0 \in S(a,b)$ if for each neighbourhood W_S of $\pi(s_0)$, there is a neighbourhood V_S of s_0 such that $\pi(V_S) \subset W_S$. We shall fix a neighbourhood W_S of $\pi(s)$ and determine a neighbourhood V_S of $\pi(s_0)$ that meets the required conditions.

Choose a point $s_0 \in S(a,b)$ and let $s = \pi(s_0)$. Let $x, y \in U$ be points such that $s \in I(x,y)$. Then $I(x,y) \cap S(a,c)$ is an open subset of S containing the point s (see Fig. 5.8).

Let $p \in I(x,y) \cap S(a,c)$. Form the sets

$$T^{\pm} = \left\{ \bigcap_p (C_p^{\pm} \cap U) | p \in I(x,y) \cap S(a,c) \right\} . \tag{5.7}$$

The sets $T^{\pm}$ are nonempty, because there exist points $\alpha, \gamma \in U$ such that

$$\alpha \ll a \ll c \ll \gamma ,$$

and $\alpha \in T^-, \gamma \in T^+$ by construction. Now

$$u \in T^- \Rightarrow u \ll p \,\forall\, p \in I(x,y) \cap S(a,c) ,$$

$$v \in T^+ \Rightarrow v \gg p \,\forall\, p \in I(x,y) \cap S(a,c)$$

and therefore

$$u \in T^-, v \in T^+ \Rightarrow I(u,v) \supset I(x,y) \cap S(a,c) . \tag{5.8}$$

Define

$$W = \text{int} \left\{ \bigcap_{u,v} I(u,v) | u \in T^-, v \in T^+ \right\} . \tag{5.9}$$

Owing to (5.8), W is nonempty. Therefore, by Lemma 5.4.1, W is open. Clearly,

$$W \cap S(a,c) \supset I(x,y) \cap S(a,c) . \tag{5.10}$$

But, as $x \in T^-$, $y \in T^+$, the interval $I(x,y)$ is one of the $I(u,v)$ on the right-hand side of (5.9), and therefore

$$W \cap S(a,c) \subset I(x,y) \cap S(a,c) . \tag{5.11}$$

Combining (5.10) and (5.11), we have

$$W_S = W \cap S(a,c) = I(x,y) \cap S(a,c) . \tag{5.12}$$

Next, choose $x_0, y_0 \in U$ such that $x_0 \ll s_0 \ll y_0$ and $V_S = I(x_0,y_0) \cap S(a,b) \subset \pi^{-1}(W_S)$. (See Fig. 5.8; for pictorial clarity, the points x_0 and s_0 are not labelled in the figure, but the point s_0 is marked by a black dot.) Then V_S is a neighbourhood of s_0 in $S(a,b)$, and

$$\pi(V_S) \subset W_S \ ,$$

which proves the continuity of π. The continuity of π^{-1} is proven in exactly the same manner. ∎

Proposition 5.4.3 *Let U be a D-set, $x, y \in U, x \ll y$. Let*

$$x \ll a \ll b \ll y \ ,$$

$$x \ll a' \ll b' \ll y \ .$$

Then

$$S(a, b) \overset{\text{hom}}{=} S(a', b') \ .$$

Proof: Clearly, $a, b, a', b' \in U$.

1. Consider the triple x, a, y. Here $a \in I(x, y)$, and therefore, from Proposition 5.4.2,

$$S(x, y) \overset{\text{hom}}{=} S(a, y) \ . \tag{5.13}$$

2. Now consider the triple a, b, y. Here $b \in I(a, y)$, and therefore, again from Proposition 5.4.2,

$$S(a, b) \overset{\text{hom}}{=} S(a, y) \ . \tag{5.14}$$

It follows from (5.13) and (5.14) that

$$S(a, b) \overset{\text{hom}}{=} S(x, y) \ . \tag{5.15}$$

Replacing a, b by a', b' respectively, one has

$$S(a', b') \overset{\text{hom}}{=} S(x, y) \ . \tag{5.16}$$

The result follows from (5.15) and (5.16). ∎

A D-set in not necessarily an order interval; but, since the family of open order intervals contained in D-sets is a base for the order topology, a D-set can be covered by open order intervals. Consequently, Proposition 5.4.3 can be improved as follows:

Theorem 5.4.4 *Let U be an l-connected D-set, $a, b, a', b' \in U$, $a \ll b$ and $a' \ll b'$. Then*

$$S(a, b) \overset{\text{hom}}{=} S(a', b') \ .$$

Proof: The straightforward details are omitted. ∎

Our final result in this direction is the following:

Theorem 5.4.5 (Second homogeneity property: $S(a,b)$**)** *Suppose that every light ray in M has an overlapping cover, and let* $U, V \subset M$ *be D-sets. Let* $a, b \in U$, $a \ll b$, $p, q \in V$, $p \ll q$, *so that* $S(a,b) \subset U$ *and* $S(p,q) \subset V$ *are spacelike hyperspheres defined by* (5.6). *Then*

$$S(a,b) \overset{\text{hom}}{=} S(p,q) .$$

Proof: Let $x \in U$ and $y \in V$. Since M is l-connected, there is an l-polygon connecting x with y, and this l-polygon has an overlapping cover. We may, without loss of generality, assume that the initial and terminal elements of this cover are U and V respectively. The result follows by repeated application of Theorem 5.4.4. ∎

6

Ordered Spaces and Complete Uniformizability

The order structure on M defined in Chaps. 3 and 4 determines not just a topology but also a family of uniformities on M, and a uniform structure admits a notion of completeness. The completed space can carry mathematical structures (such as differentiable or analytic structures) that cannot be carried by ordered spaces like $\mathbb{Q}^2$. These structures were first elaborated in the context of metric spaces, but ordered spaces may be uniformizable without being metrizable. In this chapter, we shall review the completion of uniformities determined by the order structure, define a notion of *order uniformity* and its completion, and analyse the problem of extending the order to the completed space. This will lead us to a concept which we shall call *order completion* (and which will be slightly different from uniform completion), and it will be clear from the definition that every ordered space has an order completion.

Although uniform structures are seldom used explicitly in mathematical physics, the standard notions of uniform continuity and uniform convergence are familiar to all. These notions suffice for an overview of the main argument, which is given below. A fairly detailed summary of the relevant parts of the theory of uniformities is given in Appendix A, together with references.

6.1 General Discussion and Main Results

The concept of a *uniform structure* or *uniformity* is an abstraction from the concepts of uniform continuity and uniform convergence. Uniformities as autonomous mathematical entities were first defined by André Weil [109] who, motivated by the example of topological groups, was looking for a structure that was intermediate between topological and metric structures. Uniform structures are strictly weaker than metric structures, in the sense that a metric induces a unique uniformity, but different metrics may induce the same uniformity. Similarly, uniform structures are strictly stronger than topologies, in that a uniformity induces a unique topology, but different uniformities may induce the same topology. In the theory of uniformities, the question "when

H.-J. Borchers and R.N. Sen: *Mathematical Implications of Einstein–Weyl Causality*,
Lect. Notes Phys. **709**, 67–94 (2006)
DOI 10.1007/3-540-37681-X_6

is a topological space metrizable" is replaced by the question "when is a topological space uniformizable". Like metric spaces, uniform spaces provide a framework for both geometry and functional analysis, and much of the interest in uniform spaces derives from the fact that this framework is less stringent than the one provided by metric spaces.

6.1.1 Uniformizability

As stated in the previous paragraph, a uniformity on X induces a unique topology on X.[1] This topology is called the topology of the uniformity; if only one uniformity is being considered, it is called the *uniform topology*.

Definition 6.1.1 (Uniformizability)

A topological space $(X, \mathcal{T})$ *is called* uniformizable *if there is a uniformity* $\mathcal{E}$ *on X that induces the topology* $\mathcal{T}$ *on it.*

It is a standard result in the theory of uniformities that a topological space is uniformizable if and only if it is Tychonoff [130]. Therefore, in view of our earlier result that every ordered space is Tychonoff (Corollary 4.6.3), we have the *fundamental theorem on the uniformizability of ordered spaces*:

Theorem 6.1.2 (Uniformizability theorem for ordered spaces)

Every ordered space is uniformizable.   ∎

Terminology 6.1.3 In what follows, we shall adopt the convention that, in the context of ordered spaces, the term *uniformity* will always mean one that has the order topology as its uniform topology. The discrete and trivial uniformities (Appendix A, Examples A.1.6), for example, are not of interest to us.

6.1.2 Completeness and Complete Uniformizability

The notion of completeness for uniform spaces is defined via Cauchy nets or Cauchy filters.[2] A uniform space is called *complete* if (and only if) every Cauchy filter in it converges. A uniform space X that is not complete may be completed by adjoining "ideal points" to make every Cauchy filter converge; the completion is the space of Cauchy nets or filters in the original space. The completed space $\widetilde{X}$ has a uniform structure, and X is densely, and uniformly, embedded in $\widetilde{X}$ (see Appendix A).

---

[1] The uniform topology is defined explicitly, and necessary and sufficient conditions for it to be Hausdorff stated, in Appendix A, Sect. A.3. The main results on the uniformizability of topological spaces and (for comparison) the metrizability of uniform spaces are given in Sect. A.5.

[2] In spaces satisfying the first axiom of countability, Cauchy sequences would suffice to define the notion of completeness.

A topological space is called *completely uniformizable* if there exists a uniformity in which the space is complete, and which induces its topology. It is natural to ask when a topological space is completely uniformizable. The answer, which was provided by Shirota in 1952 [98], involves a set-theoretic subtlety which is briefly discussed in Appendix A. In the framework for mathematics provided by the Zermelo-Fraenkel axioms for set theory plus the axiom of choice,[3] this subtlety becomes irrelevant, and Shirota's theorem reduces to the following:

Theorem 6.1.4 *A topological space X is completely uniformizable iff*

a) *X is a Tychonoff space.*

b) *X is a closed subspace of a product of real lines.*

Since the uniform completion of an ordered space is completely uniformizable by definition, we have the following result:

Theorem 6.1.5 *The uniform completion of an ordered space is, in its uniform topology, a closed subspace of a product of real lines.*

This theorem may justly be regarded as the key to further mathematical structures on order complete spaces.

In view of Theorems 6.1.2 and 6.1.5, interest shifts naturally to the following question: Let $\widetilde{M}$ be a uniform completion of M. Can the order structure M be extended to $\widetilde{M}$? This question, or rather an appropriate modification of it, will occupy us for the rest of this chapter, following the definition of the order uniformity on D-sets.

6.2 Order and Uniformization

If a uniformizable space is compact Hausdorff then, according to Theorem A.3.4, its uniformization is unique. This may no longer be true if the space is not compact Hausdorff. An ordered space M is always Hausdorff but never compact, and may admit of inequivalent uniformizations. However, this turns out *not* to be a problem. We begin with the definition of the *order uniformity* on D-sets.

6.2.1 The Order Uniformity on D-sets

We shall define the order uniformity on D-sets as a diagonal uniformity (Def. A.1.5), via a filter base.

---

[3] Denoted ZFC; see footnote 6 on p. 165

Lemma 6.2.1 *Let M be an ordered space, $D \subset M$ a D-set in it, and let $\{D^\alpha | \alpha \in A\}$ be a D-base for the relativized order topology on D. Denote by $\mathcal{A}$ the family of all D-bases for D, choose a particular $A \in \mathcal{A}$, and define*

$$E^A = \bigcup_{\alpha \in A} D^\alpha \times D^\alpha. \tag{6.1}$$

Then the family $\{E^A | A \in \mathcal{A}\}$ is a base for a filter $\mathcal{E}_D$ on $U \times U$.

Proof: We need to prove that if $E^1, E^2 \in \mathcal{E}_D$, $E^1 \cap E^2 \neq \emptyset$, then there exists $E^3 \in \mathcal{E}_D$ such that $E^3 \subset E^1 \cap E^2$. This is an immediate consequence of the corresponding result on bases for topologies. ∎

The result we seek is the following:

Theorem 6.2.2 *The filter $\mathcal{E}_D$ of Def. 6.2.1 defines a diagonal uniformity on D.*

Proof: We have to verify the conditions a)–c) of Def. A.1.5.

Condition a), if $E \in \mathcal{E}_D$ then $\Delta \subset E$, follows immediately from the definition. Condition b), if $E \in \mathcal{E}_D$ then there exists an entourage F such that $F \in E^{-1}$, follows from the fact that every entourage E is a superset of some symmetric entourage F, which is a subset of E^{-1} if it is a subset of E. Condition c) states that for any entourage E, there exists an entourage F such that $F \circ F \subset E$. This is true because $F = D \times D$ is itself an entourage. ∎

The following result is easy to prove, but fundamental:

Theorem 6.2.3 *The order uniformity $\mathcal{E}_D$ on D is Hausdorff.*

Proof: It is enough to prove that, for any two distinct points $x, y \in D$, there exists an entourage E that does not contain the pair (x, y). Fix x and y, $x \neq y$. Since one-point sets are closed in the order topology, there exists a D-cover $\mathcal{D}^{(-)}$ of $U \setminus \{x\}$. Let D_x be a D-set such that $x \in D_x$ but $y \notin D_x$. Then $\mathcal{D} = \mathcal{D}^{(-)} \cup \{D_x\}$ is a D-cover of M such that $D^\alpha \in \mathcal{D}$ implies that D^α does not contain both x and y. Then the entourage $E = \cup(D^\alpha \times D^\alpha)$, where the union is over all $D^\alpha \in \mathcal{D}$, does not contain the pair (x, y). ∎

6.2.1.1 The Topology of the Order Uniformity on D

Recall that for $x \in X$ and a relation R on X, the set $R[x]$ is defined as

$$R[x] = \{y \,|\, (x, y) \in R\}.$$

It follows that if R is an entourage, then $R[x]$ contains a D-set $D_x^\alpha \ni x$. Let $\mathcal{K}$ be the family of relations on U. Then the family of sets

$$\{R[x] \,|\, x \text{ fixed}, \ R \in \mathcal{K}\},$$

which is a system of neighbourhoods of x in the uniform topology, is also a system of neighbourhoods of x in the order topology. That is:

Theorem 6.2.4 *The topology of the order uniformity $\mathcal{E}_D$ on D is the same as the order topology on D.*   ∎

6.2.1.2 The Relativization Lemma for the Order Uniformity

Lemma 6.2.5 *Let M be an ordered space and $D_1, D_2 \subset M$ be D-sets such that $D_3 = D_1 \cap D_2$ is nonempty. Then the relativizations of the order uniformities on D_1, D_2 to D_3 are identical with the order uniformity on D_3.*

Proof: Follows immediately from Proposition 4.2.16, which is that the intersection of two D-sets is a D-set.   ∎

This lemma will be called *the relativization lemma*. It will be used extensively in the following, often without explicit mention of the source.

6.3 Uniform Completions of Ordered Spaces

Not being compact, an ordered space will generally admit of inequivalent uniformizations, which in turn will lead to different (uniform) completions. From our point of view, a completion will be relevant only if the order can be extended to the completed space. There exist nontrivial uniform completions to which the order of M *cannot* be extended. A good "universal" example is βM, the Stone-Čech compactification of a Tychonoff space M; βM is uniformly complete in the pseudometric uniformity (Theorem A.8.1); according to our definition of order, a compact space cannot be ordered.

Additionally, completions of M may result in obstructions other than compactness to extending the order of M. For instance, they may introduce boundaries,[4] and spaces with boundaries cannot be ordered in our sense. Alternatively, they may fill in "missing sets" such as punctures or cuts. We illustrate these possibilities below by simple examples of subspaces of the Minkowski plane. In all these examples the uniformity is the metric uniformity, the metric is the Euclidean metric and uniform completion coincides with the metric completion.

Examples 6.3.1

1. $X = \{(x, y); x \in \mathbb{R}, y > 0\}$. The completion $\widetilde{X}$ of X is the half-space $\widetilde{X} = \{(x; y); x \in \mathbb{R}, y \geq 0\}$, which is a manifold with boundary.

2. $X = \mathbb{R}^2 \setminus \bar{B}_O$, where $\overline{B}_O$ is the closed unit disc with centre at O. The completion of X is $\widetilde{X} = \mathbb{R}^2 \setminus B_O$, where B_O is the *open* unit disc with centre at O.

---

[4] If a differentiable manifold is not complete, its completion may be a manifold with boundary. The open disk in $\mathbb{R}^2$ provides a simple example.

3. The punctured plane $X = \mathbb{R}^2 \setminus \{O\}$, where O is the origin of coordinates. The completion of X is $\mathbb{R}^2$.

4. The cut plane $X \setminus \{(x, y); 0 \leq x < \infty, y = 0\}$. Again, the completion of X is $\mathbb{R}^2$.

In cases 1) and 2), the completed space cannot be ordered; light rays do not have end points. In cases 3) and 4), completion fills in the "missing sets", which are the puncture in case 3), and the cut in case 4). The order of X *can* be extended to the completed space, but at the cost of joining pairs of light rays in X into one in $\widetilde{X}$; in case 3), there are two such pairs, namely the light rays $x = y$ in the first and third quadrants of X, and the rays $x = -y$ in the second and fourth quadrants. In case 4), every light ray $x = y + a$, $a \geq 0$ in $\widetilde{X}$ joins together two light rays in X. We would like to avoid such situations. Moreover, the concept of completion of ordered spaces should be invariant under order equivalence transformations (Sect. 4.7).

6.4 The Concept of Order Completion

Suppose that one is given a completion (or some modification thereof) of M that can support an order structure. One has then to extend the order structure from M to this completion. This is a constructive procedure, and we shall gain further insight by considering, briefly, the nuts and bolts of the process.

In view of the way that order was defined, it would be reasonable to regard the extension of order to the new space as a two-step process. At the first step, one will have to complete the "old" light rays and define the new ones. At the second step, one will have to verify that the totality of light rays in the new space satisfies the order axioms. Recall now that these axioms were stated in terms of light rays *and their intersections*, and that we have effective control over these intersections, in the original space, only in D-sets. It therefore makes sense to try to extend the order D-set by D-set, using overlapping covers (Assumption 5.3.3). To do so it should suffice to have a localized version of the notion of completion. For this it is enough to have the order uniformity on D-sets, which was defined in Sect. 6.2.1. Under the assumption that follows, the order uniformity is the unique uniformity on a D-set that is compatible with its topology. The relativization Lemma 6.2.5 will be the tool that will enable us to extend the order, D-set by D-set, to the new space.

Assumption 6.4.1 (Local precompactness assumption)

From now on (unless the contrary is stated explicitly) all ordered spaces M will be assumed to be locally precompact. That is, the uniform subspace $(D, \mathcal{E}_D)$ will, by assumption, be totally bounded for every D-set $D \subset M$. Here $\mathcal{E}_D$ is the order uniformity on D. $\qquad\square$

It follows immediately from this assumption that the uniform completion of $(D, \mathcal{E}_D)$, which we shall denote by $\widetilde{D}$, is compact (Theorem A.6.12). By Theorem A.7.2, $\widetilde{D}$ is Hausdorff. Therefore, by Theorem A.7.4, it possesses a unique uniformity that is compatible with its topology. It is clear that the order uniformity on D is precisely the relativization of this uniformity to D.

Let D be a D-set in M, and define

$$\check{D} = \text{int } \widetilde{D} . \tag{6.2}$$

Let now $\{D^\alpha | \alpha \in M\}$ be a D-cover of M, and define

$$\check{M} = \bigcup_{\alpha \in A} \check{D}^\alpha . \tag{6.3}$$

Then $\check{D}^\alpha \subset \check{M} \ \forall \ \alpha \in A$, and the family of subsets $\{\check{D}^\alpha | \alpha \in A\}$ is a base for a topology on $\check{M}$. We shall furnish $\check{M}$ with this topology. The sets $\check{D}$ will be called $\check{D}$-sets in $\check{M}$; the terminology will be justified later. Finally, define

$$\widetilde{M} = \bigcup_{\alpha \in A} \widetilde{D}^\alpha . \tag{6.4}$$

Definition 6.4.2 (Topology on $\widetilde{M}$)

The topology $\mathcal{T}$ on $\widetilde{M}$ is defined by taking the closure of the family $\{\widetilde{D}^\alpha\}$ under finite unions and arbitrary intersections as its closed sets. □

$\mathcal{T}$ is indeed a topology, because $\{\widetilde{D}^\alpha | \alpha \in A\}$ covers $\widetilde{M}$. The sets $\check{D}^\alpha$ are open in the topology $\mathcal{T}$. It is the only topology that we shall ever consider on $\widetilde{M}$. Note that $\check{M} \subset \text{int } \widetilde{M} \subset \widetilde{M}$ and $\widetilde{M} = \text{cl } \check{M}$.

Definition 6.4.3 (Order completion)

Let M be an ordered space with a locally precompact uniformity. The *order completion* of M will be defined to be the space $\check{M}$ of (6.3), with the topology defined by the family of $\check{D}$-sets (6.2) as base. □

It is clear from the Definition 6.3 that $\check{M}$ does not have boundaries. Nor does it fill in the missing sets of M, as D-sets, by definition, "avoid" the missing sets. In Example 3 of 6.3.1, no D-set in M includes the puncture, and in Example 4, no D-set in M crosses the cut. In general, $\check{M} \subset \text{int } \widetilde{M}$, the equality $\check{M} = \text{int } \widetilde{M}$ holding if and only if M does not have cuts or punctures.

Notations 6.4.4

We begin with setting up some notations. In the following, D will denote a D-set in M.

1. We shall denote by $\tilde{l}$ the closure, in $\tilde{M}$, of the light ray $l \subset M$, and set $\check{l} = \tilde{l} \cap \check{M}$. Since l is linearly ordered and satisfies the density Axiom 3.1.2(b), its closure is locally the same as its Dedekind completion. Therefore $\check{l}$ is ordered, and is locally homeomorphic with $\mathbb{R}$. We shall use the same symbols $<^l$ and $<^{ll}$ to denote the orders on $\check{l}$. For $x, y \in \check{l}$, $x <^l y$, we shall denote the open and closed segments between x and y by $\check{l}(x, y)$ and $\check{l}[x, y]$ respectively.

2. Let $A \subset D$. We shall denote the uniform completion of A in $(D, \mathcal{E}_D)$ by $\tilde{A}$. Then, by Theorem A.7.5, $\tilde{A}$ is a closed subset of $\tilde{D}$. It is therefore a closed subset of $\tilde{M}$.

3. The sets $\check{D} \subset \check{M}$ were defined by (6.2). (We shall prove, eventually, that a $\check{D}$-set is a D-set in $\check{M}$.)

4. Let A be a closed subset of D. We define: $\check{A} = \tilde{A} \cap \check{D}$. That is, if A is closed in D, then $\check{A}$ is closed in $\check{D}$.

5. Let U be an open subset of D. We define: $\check{U} = (\text{int } \tilde{U}) \cap \check{D}$. That is, if U is open in D, then $\check{U}$ is open in $\check{D}$. $\check{U}$ is also the order completion of the ordered space U. $\qquad\square$

Upon extension of the order from M to $\check{M}$, it will be seen that these notations are consistent with the ones that we have been using so far.

Definition 6.4.5 (Local Cones and D-intervals)

Let M be an ordered space, $a \in D \subset M$ where D is a D-set, and $C_a^\pm$ the forward and backward cones at a. We define

$$C_{a;D}^\pm = C_a^\pm \cap D , \tag{6.5}$$

and call them *local cones* at a. By Theorem 4.5.4, the relations

$$\beta C_{a;D}^\pm = \partial C_{a;D}^\pm \cap D \tag{6.6}$$

always hold for local cones. Define successively

$$C_{a;\check{D}}^\pm = \text{considered as a subset of } C_{a;D}^\pm \text{ in } \check{D};$$

$$\tilde{C}_{a;\check{D}}^\pm = \text{the uniform completion of } C_{a;\check{D}}^\pm;$$

$$\check{C}_{a;\check{D}}^\pm = \tilde{C}_{a;\check{D}}^\pm \cap \check{D}; \tag{6.7}$$

$$\tau \check{C}_{a;\check{D}}^\pm = \text{int } \check{C}_{a;\check{D}}^\pm. \tag{6.8}$$

$\tilde{C}_{a;\check{D}}^\pm$, the uniform completion of $C_{a;\check{D}}^\pm$, has topological boundaries; the mantle and the base of the cone. We wish to keep the mantle but excise the base, and this is achieved by taking its intersection with $\check{D}$ (see (6.7)).

Let $\eta \in \check{M}$ be a point introduced by the order completion. The order structure of M, the topology of $\check{M}$ and the completeness of the $\tilde{D}$ make it possible to define *local cones* at η before light rays through η are defined, and these definitions are given below.

Let $D \subset M$ be a D-set such that $\eta \in \check{D}$. Then there exists a filter base of closed order intervals $I[r_\alpha, s_\alpha]$ in D that converges to $\eta \in \check{M}$, and we define the local cone $\check{C}^+_{\eta;\check{D}}$ as

$$\check{C}^+_{\eta;\check{D}} = \left(\bigcap_{r_\alpha} \widetilde{C}^+_{r_\alpha;\check{D}} \right) \cap \check{D} . \tag{6.9}$$

$\check{C}^-_{\eta;\check{D}}$ would then be defined as

$$\check{C}^-_{\eta;\check{D}} = \left(\bigcap_{s_\alpha} \widetilde{C}^-_{s_\alpha;\check{D}} \right) \cap \check{D} . \tag{6.10}$$

Next, if $x, y \in \check{C}^+_{x;\check{D}}$, we define

$$\begin{aligned} \check{I}[x, y] &= \check{C}^+_{x;\check{D}} \cap \check{C}^-_{y;\check{D}} , \\ \check{I}(x, y) &= \tau\check{C}^+_{x;\check{D}} \cap \tau\check{C}^-_{y;\check{D}} \end{aligned} \tag{6.11}$$

and call them $\check{D}$-intervals. □

The definitions of local cones and $\check{D}$-intervals given above are purely topological. For the moment, a cone mantle is available to us only locally, and only as part of the topological boundary of a local cone. As already remarked, the topological boundary of $\widetilde{C}^\pm_{x;\check{D}}$ contains the base as well as the mantle. In (6.7), the base was excised by intersecting $\widetilde{C}^\pm_{a;\check{D}}$ with $\check{D}$. In the same spirit, we define the *mantle operator* $\check{\partial}$ as follows:

Definition 6.4.6 (Mantle operator)

$$\check{\partial}\check{C}^\pm_{x;\check{D}} = (\partial\check{C}^\pm_{x;\check{D}}) \cap \check{D} , \tag{6.12}$$

where ∂ is the standard topological boundary operator. □

Finally, we extend the definition (5.6) of spacelike hyperspheres in D-sets to their order completions $\check{D}$:

Definition 6.4.7

$$\check{S}(x, y) = \check{\partial}\check{C}^+_{x;\check{D}} \cap \check{\partial}\check{C}^-_{y;\check{D}} . \tag{□}$$

Observe that if $x, y \in M$, then $\check{S}(x, y)$ is the completion of $S(x, y)$ in its relative uniformity.

The notations and terminology in the Definitions (6.7)–(6.11) remain to be justified. When this is done, the Notation 6.4.7 will automatically be justified.

6.4.1 Notations Special to This Chapter

In the rest of this chapter, we shall use a few special notations, defined below:

1. i) Points in M will be denoted by lower-case Latin letters, *excepting* the letters x, y, z.

   ii) The letters x, y, z will denote points in $\check{M}$ which may or may not be in M.

   iii) Points in $\check{M}$ which are *not* in M will be denoted by the lower-case Greek letters ξ, η, ζ.

2. ℓ will denote a new light ray in $\check{M}$, i.e., one which is not the completion of a ray l in M.

6.5 Some Basic Results

We shall now establish some basic results concerning the objects defined above, using the completeness of $\widetilde{M}$, the topology of $\check{M}$ and the order structure of M. These results will greatly facilitate the extension of order from M to $\check{M}$. However, we shall establish them under a restrictive assumption, which is:

Assumption 6.5.1 (First countability assumption)

From now on, all ordered spaces M will be assumed to satisfy the first axiom of countability, unless the contrary is stated explicitly.

The fact that sequential convergence suffices to describe the topology of a first countable space will be fully exploited the subsequent developments.

6.5.1 Symmetry Properties

We extend the Definition 3.2.6 of the order "$<$" as follows:

Definition 6.5.2

$$1)\ x < y \text{ iff } y \in \check{C}^{+}_{x;\check{D}}\ ;$$
$$2)\ x > y \text{ iff } y \in \check{C}^{-}_{x;\check{D}}\ . \qquad \square$$

In M, the equivalence $y \in C^{+}_{x} \Leftrightarrow x \in C^{-}_{y}$ (3.2.5) follows immediately from the polygon lemma, which is not yet available in $\check{M}$. However, the result – which is both order-theoretic and topological – can be proved by purely topological means. The proof is given in Lemmas 6.5.4 and 6.5.5. We begin with a preliminary lemma.

Lemma 6.5.3 *Let $\xi \in \check{D}$. Then there exist sequences $\{a_n\}, \{b_n\}, a_n, b_n \in D \ \forall \ n \in \mathbb{N}^{+}, a_n \ll a_{n+1}$ and $b_n \gg b_{n+1}$ that converge to ξ from below and from above respectively.*

Proof: By definition, ξ is the limit of some Cauchy sequence $\{c_n\}, c_n \in D \; \forall \; n \in \mathbb{N}^+$. We may assume, without loss of generality, that $c_i \neq c_j$ for $i \neq j$.

1. There exists a neighbourhood $U_1 \subset \check{D}$ of ξ such that $c_1 \notin U_1$. Choose $a_1, b_1 \in U_1$ such that $\xi \in \check{I}(a_1, b_1)$. Then $\{c_n\}$ is ultimately in $\check{I}(a_1, b_1)$, and the intersection $\{c_n\} \cap \check{I}(a_1, b_1)$ contains a Cauchy subsequence of $\{c_n\}$. Re-label this subsequence as $\{c_n\}$, starting with $n = 2$.

2. There exists a neighbourhood $U_2 \subset \check{I}(a_1, b_1)$ of ξ such that $c_2 \notin U_2$. Choose $a_2, b_2 \in U_2$ such that $\xi \in \check{I}(a_2, b_2)$. Then $\{c_n\}$ is ultimately in $\check{I}(a_2, b_2)$, and the intersection $\{c_n\} \cap \check{I}(a_2, b_2)$ contains a Cauchy subsequence of $\{c_n\}$. Re-label this subsequence as $\{c_n\}$, starting with $n = 3$.

3. Iterating this process, we obtain a chain of intervals

$$\check{I}(a_1, b_1) \supset \check{I}(a_2, b_2) \supset \cdots \check{I}(a_n, b_n) \supset \cdots$$

such that $\xi \in \cap_n \check{I}(a_n, b_n)$. Standard arguments show that the intersection $\cap_n \check{I}(a_n, b_n)$ contains no point other than ξ, and that $a_n \to \xi$ and $b_n \to \xi$. The sequences $\{a_n\}$ and $\{b_n\}$ are the desired ones. ∎

Lemma 6.5.4 Let $x, y \in \check{D}$. If $y \in \tau \check{C}^+_{x; \check{D}}$ then $x \in \tau \check{C}^-_{y; \check{D}}$, and the same with order reversed.

Proof: Let $y \in \tau \check{C}^+_{x; \check{D}}$ and let $\{a_n\}, a_n \gg a_{n+1}$ and $\{b_n\}, b_n \gg b_{n+1}$ be sequences in D that converge in $\check{D}$ to x and y respectively. Since $x \neq y$, they may be separated by disjoint open subsets O_x and O_y of $\check{D}$, and $\{a_n\}, \{b_n\}$ will ultimately be in O_x and O_y respectively. Since $y \in \tau \check{C}^+_{x; \check{D}}$ we may, without loss of generality, assume that $y \in O_y \subset \tau \check{C}^+_{x; \check{D}}$. Then, for $m, n > N$, $a_m \ll b_n$, so that $a_m \in \check{C}^-_{b_n; \check{D}}$ for $m, n > N$. It follows from this that $x \in \tau \check{C}^-_{y; \check{D}}$. Mutatis mutandis, the same proof holds with order reversed. ∎

Lemma 6.5.5 Let $x, y \in \check{D}$. If $y \in \eth \check{C}^+_{x; \check{D}}$ then $x \in \eth \check{C}^-_{y; \check{D}}$, and the same with order reversed.

Proof: By contradiction. If $x \notin \eth \check{C}^+_{y; \check{D}}$ then $x \in \tau \check{C}^-_{y; \check{D}}$, so that, by Lemma 6.5.4, $y \in \tau \check{C}^-_{x; \check{D}}$, which contradicts the assumption. ∎

Corollary 6.5.6

$$y \notin \check{C}_{x; \check{D}} \Leftrightarrow x \notin \check{C}_{y; \check{D}} .$$

Proof: Follows immediately from Lemmas 6.5.4 and 6.5.5: ∎

Notations and Terminology 6.5.7

In view of the above results, we shall use the notations $x \ll y$, $y \gg x$ and $x < y$, $y > x$, wherever appropriate, in $\check{M}$. We shall also say that x and y are spacelike to each other if $y \notin \check{C}_{x; \check{D}}$. Once again, the terminology will be justified upon extension of the order from M to $\check{M}$.

6.5.2 Separation Theorems

Next, we establish a few basic separation theorems. These are nontrivial because the order of M has not yet been extended to $\check{M}$. The proofs involve a subtle interplay between the order completeness and topology of $\check{M}$, and the order structure of M.

Lemma 6.5.8 *Let $D \subset M$ be a D-set, and the points $r, s, t, u \in D$ such that*

$$r \ll s \ll t \ll u .$$

Then

$$\check{I}[s,t] \subset \check{I}(r,u) .$$

Proof:

1. Every point in $\check{I}[s,t]$ is defined by a Cauchy sequence lying wholly in $I[s,t]$, and therefore in $I[r,u]$. It follows that $\check{I}[s,t] \subset \check{I}[r,u]$.
2. Let $a \in \beta I[r,u] \subset \partial \check{I}[r,u]$. Then $a \notin \check{I}[s,t]$, because $a \notin \check{I}[s,t] \cap D = I[s,t]$. Since $\check{D}$ is regular, there exists an open set $O_a \subset \check{D}_a$ such that $a \in O_a$ and $O_a \cap \check{I}[s,t] = \emptyset$. Let

$$O = \bigcup_{a \in \beta I[r,u]} O_a$$

Then O is disjoint from $\check{I}[s,t]$. Since $\beta I[r,u]$ is dense in $\partial \check{I}[r,u]$, O covers $\partial \check{I}[r,u]$. Therefore $\partial \check{I}[r,u] \cap \check{I}[s,t] = \emptyset$, which proves the desired result. ∎

Notations 6.5.9 If O is an open set in $S(a,b) \subset D$, we shall denote its closure in $S(a,b)$ by $\overline{O}$, its completion by $\widetilde{O}$ and the interior of $\widetilde{O}$ by $\check{O}$. Then $\widetilde{O}, \check{O} \subset \check{S}(a,b)$.

Lemma 6.5.10 *Let $D \subset M$ be a D-set, $S(a,b)$ a spacelike hypersphere in D and $\check{S}(a,b)$ its uniform completion. If V, W are open sets in $S(a,b)$ such that $\overline{W} \subset V$, then*

$$\widetilde{W} \subset \check{V} .$$

Proof: Mutatis mutandis, the proof of Lemma 6.5.8 applies here. ∎

Theorem 6.5.11 *Let $U \subset M$ be a D-set and $a, b, a', b' \in U$ such that $a \ll b$, $a' \ll b'$ and $I(a,b) \cap I(a',b') = \emptyset$. Then*

$$\check{I}(a,b) \cap \check{I}(a',b') = \emptyset .$$

Proof: For any $x \in \check{I}(a,b)$, there exist $r_x, s_x \in I(a,b)$, $r_x \ll s_x$ such that $x \in \check{I}(r_x, s_x)$. By Lemma 6.5.8, $\check{I}[r_x, s_x] \subset \check{I}(a,b)$, and therefore $\check{I}(r_x, s_x) \subset \check{I}(a,b)$. Then

$$\check{I}(a,b) = \bigcup_{x \in \check{I}(a,b)} \check{I}(r_x, s_x) .$$

Similarly,

$$\check{I}(a',b') = \bigcup_{x' \in \check{I}(a',b')} \check{I}(r_{x'}, s_{x'})$$

where $x' \in \check{I}(r_{x'}, s_{x'})$. Suppose now that there exist $x \in \check{I}(a,b)$, $x' \in \check{I}(a',b')$ such that $x = x'$. Then there exist $r_x, s_x \in I(a,b)$, $r_{x'}, s_{x'} \in I(a',b')$ such that $x \in \check{I}(r_x, s_x)$, $x' \in \check{I}(r_{x'}, s_{x'})$. Then $x = x' \Rightarrow I(r_x, s_x) \cap I(r_{x'}, s_{x'}) \neq \emptyset$, which contradicts the assumption that $I(a,b) \cap I(a',b') = \emptyset$. ∎

Theorem 6.5.12 *Let W, W' be open sets in $S(a,b)$ such that $W \cap W' = \emptyset$. Then $\check{W} \cap \check{W}' = \emptyset$.*

Proof: Let $x \in \check{W}$, $x' \in \check{W}'$ such that $x = x'$. Since $x \in \check{W}$, $x' \in \check{W}'$, there exist sequences $\{c_n\}$, $c_n \in W$ and $\{c'_n\}$, $c'_n \in W'$ that converge to $x \in \check{W}$ and $x' \in \check{W}'$ respectively. Let $\check{V}_x, \check{V}'_{x'}$ be open subsets of $\check{W}, \check{W}'$ such that $x \in \widetilde{V}_x \subset \check{W}_x$ and $x' \in \widetilde{V}'_{x'} \subset \check{W}'_{x'}$. Let $V_x \in \check{V}_x \cap W$ and $V_{x'} \in \check{V}'_{x'} \cap W'$. Then $V_x \cap V'_{x'} = \emptyset$. Furthermore, $\{c_n\}$ is ultimately in V_x and $\{c'_n\}$ is ultimately in $V'_{x'}$. As $x = x'$, the sequences $\{c_n\}$ and $\{c'_n\}$ are both ultimately in $V_x \cap V'_{x'}$, which is a contradiction. ∎

Theorem 6.5.13 *Let $D \subset M$ be a D-set and p, q a pair of spacelike points in D, i.e., $C^+_{p;D} \cap C^-_{q;D} = \emptyset$. Then $\check{C}^+_{p;\check{D}} \cap \check{C}^-_{q;\check{D}} = \emptyset$.*

Proof: There exist points $r, s \in D$ such that $r \ll p$, $q \ll s$ and r, s are spacelike to each other, i.e.,

$$C^+_{p;D} \subset C^+_{r;D}, \quad C^-_{q;D} \subset C^-_{s;D}$$

and

$$C^+_{r;D} \cap C^-_{s;D} = \emptyset .$$

It follows from Theorem 6.5.11 that if $x \in \tau \check{C}^+_{r;\check{D}}$ and $y \in \tau \check{C}^-_{s;\check{D}}$ then $x \neq y$. The result follows from the facts that $\check{C}^+_{p;\check{D}} \subset \tau \check{C}^+_{r;\check{D}}$ and $\check{C}^-_{q;\check{D}} \subset \tau \check{C}^-_{s;\check{D}}$, the proofs of which are analogous to the proof of Lemma 6.5.8. ∎

6.5.3 Density Lemmas

We now establish two density lemmas:

Lemma 6.5.14 *Let $\xi, \zeta \in \check{D}$ such that $\xi \ll \zeta$. Then there exists $\eta \in \check{D}$ such that $\xi \ll \eta \ll \zeta$.*

Proof: It is a straightforward matter to find $a, b \in D \subset \check{D}$ such that $\xi \ll a \ll b \ll \zeta$. Let $l[p,q] \subset I(a,b)$, $p <^{ll} q$. By Theorem 4.3.2, there exist points c, d such that $a \ll c \ll d \ll b$. Choose $\eta \in \check{I}(c,d)$. Then, by Lemma 6.5.4, $\xi \ll a \ll \eta \ll b \ll \zeta$. ∎

Lemma 6.5.15. Let $z \in \mathring{\partial}\check{C}^+_{x;\check{D}}$, $z \neq x$. Then there exists $y \in \mathring{\partial}\check{C}^+_{x;\check{D}}$, $x \neq y$, $y \neq z$ such that $z \in \mathring{\partial}\check{C}^+_{y;\check{D}}$.

Proof: Let $\{p_n\}$, $p_n \ll p_{n+1}$ be a sequence in D that converges to x. Let $\{q_n\}$, $q_n \ll q_{n+1}$ and $\{r_n\}$, $r_n \gg r_{n+1}$ be sequences in D that converge to z from below and from above respectively, such that $x \notin \check{C}^+_{q_0,\check{D}}$. Choose $h \in \tau\check{C}^+_{x,\check{D}} \cap D$ such that $h \notin \check{C}^+_{q_0,\check{D}}$ and $h \notin \check{C}^-_{r_0,\check{D}}$.

Let $V_n = \mathring{\partial}\check{C}^-_{h,\check{D}} \cap \check{I}[p_n, r_n]$. Then V_n is closed, complete and nonempty, and $V_{n+1} \subsetneq V_n$. Therefore $\cap_n V_n$ is nonempty. Choose $y \in V$. Then $y \in \check{C}^+_{x,\check{D}}$, as $y \in \check{C}^+_{p_n,\check{D}}$ for every n. If $y \in \tau\check{C}^+_{x,\check{D}}$ then also $z \in \tau\check{C}^+_{x,\check{D}}$, which contradicts the definition of z. Therefore $y \in \mathring{\partial}\check{C}^+_{x,\check{D}}$.

By a similar argument (with x and z interchanged), $y \in \mathring{\partial}\check{C}^-_{z,\check{D}}$. ∎

6.6 Extending the Order to $\check{M}$

The order completion $\check{M}$ of M is not yet an ordered space. We shall now extend the order on M to $\check{M}$. Logically, this would appear to be a two-step process, the steps being:

1. Defining new light rays in $\check{M}$ through points $\xi \in \check{M}$, $\xi \notin M$.

2. Verifying that the order axioms are satisfied in $\check{M}$.

However, to ensure that the definitions define new light rays uniquely, we shall find it more convenient not to attempt such a clear-cut separation of the two.

6.6.1 New Light Rays in $\check{M}$

In the rest of this chapter, we shall make the following assumption:

Assumption 6.6.1 (Dimensionality assumption)

There are infinitely many light rays through any point of M.[5]

This means that the two-dimensional case is being excluded. For the two-dimensional case, the results of interest may be established in a rather more straightforward manner, and we shall omit the details.

To establish the results that follow, we shall make extensive use of the first countability Assumption 6.5.1. As explained earlier, our constructions will be local. That is, we shall define new light rays on boundaries of $\check{D}$-intervals that lie entirely in $\check{D}$-sets, and then extend these segments backwards and forwards by means of overlapping covers.

---

[5] If there are at least three different light rays through every point, then one may convince oneself that there are infinitely many different rays through every point. Since we did not want to go into the details of this question we have posed it as an assumption.

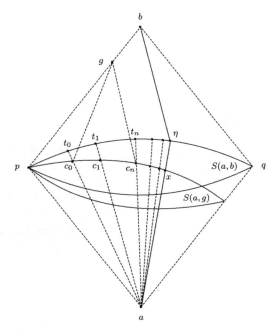

Fig. 6.1. Defining the segments $\ell[a, \eta]$ and $\ell[\eta, b]$

6.6.1.1 New Light Ray Segments on $\check{\eth}\check{C}^+_{a;\check{D}}$ and $\check{\eth}\check{C}^-_{b;\check{D}}$

Let $D \subset M$ be a D-set, $a, b \in D$ and $a \ll b$. Let $\eta \in \check{S}(a, b) \setminus S(a, b)$. Then η does not belong to any $\check{l}^+_a \cap \check{I}[a, b]$ or any $\check{l}^-_b \cap \check{I}[a, b]$. We shall define the ray segments $\ell[a, \eta]$ and $\ell[\eta, b]$ (Fig. 6.1). To do so, we shall work entirely on $\check{\eth}\check{I}[a, b]$ or on $\check{S}(a, b)$, which are closed, and therefore complete uniform subspaces of $\widetilde{M}$ that are contained in $\check{M}$. Recall that $S(a, b)$ is densely embedded in $\check{S}(a, b)$.

Construction 6.6.2 (Defining the new segments)

Let O be a nonempty open set in $S(a, b)$ such that $\eta \in \partial\widetilde{O}$, $\check{O}$ is connected and $\check{O}' = \check{S}(a, b) \setminus \widetilde{O}$ is nonempty.[6] There exists a Cauchy sequence $\{t_n\}$, $t_n \in O$, $n \in \mathbb{N}$ that converges to $\eta \in \partial\check{O}$. The light-ray segments $l[a, t_n]$ lie on $\beta I[a, b] \subset \check{\eth}\check{I}[a, b]$ for all $n \in \mathbb{N}$. Fix a point $p \in S(a, b)$ which is distinct from each of the t_n (Fig. 6.1). We shall make use of the standard maps ρ and σ that were defined in Sect. 5.1.1:

$$\rho{:}l[a, t_0] \to l[p, b]$$
$$\sigma_n{:}l[p, b] \to l[a, t_n], \quad n \in \mathbb{N}, \tag{6.13}$$

where σ_0 is the inverse of ρ.

---

[6] Construction 6.6.2 does not require either $\check{O}' \neq \emptyset$ or $\check{O} \cap \check{O}' = \emptyset$. We are preparing the ground for the proof of Proposition 6.6.3 that follows.

Let c_0 be the generic point on $l[a, t_0]$, and let $g = \rho(c_0)$ (see Fig. 6.1). Let

$$c_n = \beta C_g^- \cap l_{a, t_n} \,.\tag{6.14}$$

Then

$$c_n = \sigma_n \circ \rho(c_0) \in S(a, g) \ \forall \ n \in \mathbb{N} \,.\tag{6.15}$$

The sequence $\{c_n\}_{n \in \mathbb{N}}$ is Cauchy in $S(a, g)$ and converges in $\check{S}(a, g)$, which is proven as follows.

Let $\pi : S(a, g) \to S(a, b)$ be the stereographic projection from the vertex a. π is a homeomorphism (the proof of continuity of π and of π^{-1} was given in Sect. 5.4, Proposition 5.4.2). Now $S(a, g)$ and $S(a, b)$ are uniform subspaces of M, and it follows immediately from the definition of uniform continuity (Def. A.4.1) that π and π^{-1} are uniformly continuous. Then, from Proposition A.4.4, the fact that $\{t_n\}_{n \in \mathbb{N}}$ is Cauchy in $S(a, b)$ implies that $\{c_n\}_{n \in \mathbb{N}}$ is Cauchy in $S(a, g)$. Its convergence in $\check{S}(a, g)$ follows from the completeness of the latter.

Set now

$$\lim_{n \to \infty} c_n = x \,.\tag{6.16}$$

As c_0 runs over $l[a, t_0]$, its image c_n (6.15) runs over $l[a, t_n]$ for each $n \in \mathbb{N}$, and the c_n lie on $\beta I[a, b]$. Therefore the (ordered) set of limits

$$\ell^*[a, \eta] = \left\{ x \mid x = \lim_{n \to \infty} c_n \right\} \,,\tag{6.17}$$

which is an exact ordered copy of each of the segments $l[a, t_n]$, lies on $\check{\partial} \check{I}[a, b]$. The light-ray segment $\ell[a, \eta]$ is defined to be the Dedekind completion of $l^*[a, \eta]$. Since $\check{\partial} \check{I}[a, b]$ is complete, it follows that $\ell[a, \eta] \subset \check{\partial} \check{I}[a, b]$.

Finally, we shall prove that:

Proposition 6.6.3 *The segment $\ell[a, \eta]$ depends, not on the particularities of the sequence $\{t_n\}$, but only upon its limit η.*

Proof: As the proof is long, it is broken up into a number of shorter steps:

1. We use the open sets $O \subset S(a, b)$ and $\check{O}' \subset \check{S}(a, b)$ that were defined in the Construction 6.6.2, and set $O' = \check{O}' \cap S(a, b)$. Then O and O' are disjoint open sets in $S(a, b)$, and $\eta \in \overline{O} \cap \overline{O}'$ in $\check{S}(a, b)$. Finally, $t_n \in O \ \forall \ n \in \mathbb{N}$. We pick a sequence $\{t_n'\}$, $n \in \mathbb{N}$ such that i) $t_0' = t_0$, ii) $t_n' \in O'$ for $n \geq 1$, and iii) $\lim t_n' = \eta$.
2. Let $\ell'[a, \eta]$ be the light ray segment defined (by Construction 6.6.2) by the sequence $\{t_n'\}$. We shall assume that $\ell'[a, \eta] \neq \ell[a, \eta]$, and arrive at a contradiction.
   We denote the standard maps : $l[p, b] \to l[a, t_n']$ by σ_n':

$$\sigma_n' : l[p, b] \to l[a, t_n'], \ n \in \mathbb{N} \,,\tag{6.18}$$

and set, by analogy with (6.15),

$$c'_n = \partial C_g^- \cap l_{a,t'_n} \, , \tag{6.19}$$

where, as in (6.14) and (6.15), c_0 is the generic point on $l[a,t'_0] = l[a,t_0]$, and $\sigma'_0 = \sigma_0$. Furthermore,

$$c'_n = \sigma'_n \circ \rho(c_0) \in S(a,g) \; \forall \; n \in \mathbb{N} \, ,$$

where $g = \rho(c_0)$, as before (Fig. 6.1). If now

$$\lim_{n \to \infty} c'_n = \lim_{n \to \infty} c_n$$

for every $c_0 \in l(a,t_0)$, then

$$\ell^\star[a,\eta] = \ell'^\star[a,\eta] \, ,$$

and therefore $\ell[a,\eta] = \ell'[a,\eta]$. Conversely, if $\ell[a,\eta] \neq \ell'[a,\eta]$, then there must be a point $c_0 \in l(a,t_0)$ such that

$$\lim c_n \neq \lim c'_n \, . \tag{6.20}$$

In that case, we shall find that

$$\lim t_n \neq \lim t'_n \, .$$

3. From now on, c_0 will denote a *fixed* point on $l(a,t_0)$ for which the inequality (6.20) is assumed to hold. Correspondingly, the point $g = \rho(c_0)$ will also be fixed. We shall set

$$z = \lim c_n$$
$$z' = \lim c'_n \, . \tag{6.21}$$

4. Since $\check{M}$ is Hausdorff, z and z' can be separated by disjoint open sets $\check{U}, \check{U}'$ respectively. It follows that there exist points $r, s, r_0, s_0 \in \check{U} \cap D$ and $r', s', r'_0, s'_0 \in \check{U}' \cap D$ such that

$$r \ll r_0 \ll s_0 \ll s$$

and

$$r' \ll r'_0 \ll s'_0 \ll s'$$

with

$$z \in \check{I}(r_0, s_0), \quad z' \in \check{I}(r'_0, s'_0)$$

and

$$\check{I}(r,s) \cap \check{I}(r',s') = \emptyset \, .$$

Then $\{c_n\}$ is ultimately in $\check{I}(r_0, s_0)$ and $\{c'_n\}$ is ultimately in $\check{I}(r'_0, s'_0)$.

5. Let $V = \check{I}(r,s) \cap S(a,g)$, $V_0 = \check{I}(r_0,s_0) \cap S(a,g)$ and $V' = \check{I}(r',s') \cap S(a,g)$, $V_0' = \check{I}(r_0',s_0') \cap S(a,g)$. Then $V \cap V' = \emptyset$, $\{c_n\}$ is ultimately in V_0 and $\{c_n'\}$ is ultimately in V_0'. Furthermore, by Lemma 6.5.10, $\widetilde{V_0} \subset \check{V}$, $\widetilde{V_0'} \subset \check{V}'$.

6. Let $W = \pi(V)$, $W_0 = \pi(V_0)$ and $W' = \pi(V')$, $W_0' = \pi(V_0')$, where $\pi : S(a,g) \rightarrow S(a,b)$ is the projection homeomorphism. Then $\pi(c_n) = t_n$, $\pi(c_n') = t_n'$, W, W_0, W', W_0' are open in $S(a,b)$, $\{t_n\}$ is ultimately in W_0 and $\{t_n'\}$ is ultimately in W_0'. Furthermore, $W \cap W' = \emptyset$. By Theorem 6.5.12, the latter implies that $\check{W} \cap \check{W}' = \emptyset$. But $\lim t_n \in \widetilde{W_0}$, $\lim t_n' \in \widetilde{W_0'}$, and, by Lemma 6.5.10, $\widetilde{W_0} \subset \check{W}$, $\widetilde{W_0'} \subset \check{W}'$, so that $\widetilde{W_0} \cap \widetilde{W_0'} = \emptyset$, which in turn implies that $\lim t_n \neq \lim t_n'$. This contradiction establishes the desired result. ∎

Proposition 6.6.4 *Let ζ, η be distinct points in $\check{S}(a,b) \backslash S(a,b)$. Then $\ell[a,\eta] \cap \ell[a,\zeta] = \{a\}$.*

Proof: Assume that there exists a point $\vartheta \in \ell[a,\eta] \cap \ell[a,\zeta]$, $\vartheta \neq a$. From here the proof uses the same ideas and techniques that were used in the proof of Proposition 6.6.3, and the details are omitted. ∎

The segment $\ell[\eta, b]$ is defined in exactly the same manner.

6.6.1.2 Properties of $\ell[a,\eta]$

We shall now prove that $\ell[a,\eta]$ meets the requirements of a light ray segment on $\check{\eth}\check{C}^+_{a;\check{D}} \subset \check{M}$, that is, the limit $\ell[a,\eta]$ has the following properties:

1. $\ell[a,\eta]$ is totally ordered, by definition.

2. $\ell[a,\eta]$ is closed, because it is homeomorphic to a closed interval on the real line.

3. $\ell[a,\eta] \subset \eth\check{I}[a,b]$, by construction.

4. $\ell[a,\eta] \subset \eth\check{C}^-_{\eta;\check{D}}$.

   Proof:

   i) There exists a family of closed D-intervals $I[u_n, w_n] \subset D$ such that $u_{n+1} \ll u_n \ll w_n \ll w_{n+1}$, and $\cap_{n \in \mathbb{N}} \check{I}[u_n, w_n] = \eta$ (see Fig. 6.2). Then $\{w_n\}_{n \in \mathbb{N}}$ is a Cauchy sequence in D which converges to $\eta \in \eth\check{C}^-_{a;\check{D}}$. Hence for $t_n \in I[u_n, w_n]$,

   $$\check{l}[a, t_m] \subset \check{C}^-_{w_n;\check{D}} \quad \text{for} \ m \geq n \,.$$

   Then $\ell[a,\eta] \subset \check{C}^-_{w_n;\check{D}}$ for all $n \in \mathbb{N}$, so that $\ell[a,\eta] \subset \check{C}^-_{\eta;\check{D}}$.

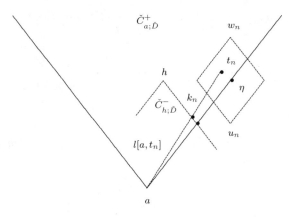

Fig. 6.2. Properties of the segment $\ell[a, \eta]$

ii) Suppose that $\zeta \in \ell[a, \eta]$ is an interior point of $\check{C}^{-}_{\eta;\check{D}}$. Then there exist $p, q \in \check{C}^{-}_{\eta;\check{D}} \cap D$ such that $\zeta \in \check{I}(p, q)$. It follows that q is spacelike to a. Therefore, by Theorem 6.5.13, $\check{C}^{+}_{a;\check{D}} \cap \check{C}^{-}_{q;\check{D}} = \emptyset$, i.e., $\zeta \notin \ell[a, \eta]$, a contradiction. $\blacksquare$

5. Let $h \in \tau C^{+}_{a;D}$ such that $u_n \notin C^{-}_{h;D}$ for $n \geq N_0$ (see Fig. 6.2). Then $\check{\partial}\check{C}^{-}_{h;\check{D}} \cap \ell[a, \eta]$ consists of a single point.[7]
   Proof: The segment $l[a, t_n]$ intersects $\beta C^{-}_{h;D}$ at a single point in M. Denote this point by k_n:
   $$k_n = l[a, t_n] \cap \beta C^{-}_{h;D} .$$
   By the same argument that led to (6.16), the sequence $\{k_n\}$ converges to a unique point $\gamma \in \check{S}(a, h)$. From the definition of $\ell[a, \eta]$ it follows immediately that $\gamma \in \ell[a, \eta]$. $\blacksquare$

6. Let $\xi \in \tau \check{C}^{+}_{a;\check{D}}$ such that $u_n \notin \check{C}^{-}_{\xi;\check{D}}$ for some n. Then $\check{\partial}\check{C}^{-}_{\xi;\check{D}} \cap \ell[a, \eta] \neq \emptyset$ consists of a single point.
   Proof: The just-established property 5) of $\ell[a, \eta]$ provides, together with the Definition 6.10 of $\check{C}^{-}_{\xi;\check{D}}$, a convergent sequence of points on $\ell[a, \eta]$. The intersection in question is the limit of this sequence. $\blacksquare$

Finally, we establish the following:

Theorem 6.6.5
$$\ell[a, \eta] = \bigcap_{n \in \mathbb{N}} \check{I}[a, w_n]$$

---

[7] Recall that we use the prefix τ to denote the interiors of cones in both M and $\check{M}$.

where the points w_n are as above, i.e., $w_n \gg w_{n+1}$ and the sequence $\{w_n\}_{n\in\mathbb{N}}$ converges to η.

Proof:

1. From the definition (6.17) of $\ell^\star[a, \eta]$ as a set of limits, it follows that

$$\ell^\star[a, \eta] \subset \check{I}[a, w_n] \ \forall \ n \in \mathbb{N} \,.$$

   Since $\check{I}[a, w_n]$ is complete and $\ell[a, \eta]$ is the completion of $\ell^\star[a, \eta]$, it follows that $\ell[a, \eta] \subset \check{I}[a, w_n]$.
2. The intersection

$$L = \bigcap_{n\in\mathbb{N}} \check{I}[a, w_n]$$

   contains no interior point of $\check{C}^+_{a;\check{D}}$. For if $z \in L$ is an interior point of $\check{C}^+_{a;\check{D}}$, then $\eta \in \check{S}(a, b) \Rightarrow \eta \notin \check{I}[z, w_1]$. Since $\check{M}$ is regular, so is $\check{D}$, and therefore there exists a neighbourhood $V_\eta \subset \check{D}$ of η which does not intersect the closed set $\check{I}[z, w_1]$. Now V_η contains every w_n for $n \geq N_0$. Therefore

$$\check{I}[z, w_1] \cap \check{I}[a, w_n] = \emptyset \ \text{ for } \ n \geq N_0 \,,$$

   which contradicts the condition $z \in L$. Therefore $L \subset \check{\partial}\check{C}^+_{a;\check{D}}$.
3. Finally, let $z \in L$ but $z \notin \ell[a, \eta]$. Since z is not an interior point of $\check{C}^+_{a;\check{D}}$, it cannot be an interior point of $\check{I}[a, w_n]$ for any n. Then necessarily

$$z \in \check{\partial}\check{I}[a, w_n] \ \forall \ n \in \mathbb{N} \,. \tag{6.22}$$

   Since $w_n \gg w_{n+1}$ for every n, the intersection $\check{\partial}C^-_{w_n;\check{D}} \cap \check{\partial}C^-_{w_{n+1};\check{D}} = \emptyset$ for every n, so that (6.22) implies that $a <^{ll} z$ (the ray segment $\ell[a, z]$ has already been defined) and $\ell[a, z] \cap \ell[a, \eta] = \{a\}$. Then

$$\check{C}^+_{z;\check{D}} \cap \ell[a, \eta] = \emptyset \,. \tag{6.23}$$

   However, $z \in L$ implies that

$$w_n \in \check{C}^+_{z;\check{D}} \ \forall \ n \in \mathbb{N} \,,$$

   which in turn implies that $\eta \in \check{C}^+_{z;\check{D}}$, which contradicts (6.23). ∎

The segment $\ell[a, \eta]$ does not lie on the completion of any light ray in M, which justifies the term *new light ray*. The proof is furnished by the following theorem:

Theorem 6.6.6 *Let $U \subset M$ be a D-set and $\ell[a, \eta] \subset \check{U}$. If $x \in \ell[a, \eta]$, $x \neq a$, then $x \notin M$.*

Proof: Suppose, to the contrary, that $x \in M$. Then there is a light ray $l_{a,x}$ through a and x in $U \subset M$, and this ray belongs to $\beta C_{a;D}^{\pm}$, i.e., $\eta \in M$, which contradicts the assumption $\eta \notin U \subset M$. ∎

Observe that Theorem 6.6.5 can be used as an alternative definition of the segment $\ell[a, \eta]$. We also have, for points in on $\ell(a, \eta)$:

Proposition 6.6.7 *Let $\xi \in \ell(a, \eta)$, and let $\{v_n\}$, $v_{n+1} \ll v_n$ be a sequence of points in $C_{a;D}^+$ that converges to ξ. Then*

$$\ell[a, \xi] = \bigcap_{n \in \mathbb{N}} \check{I}[a, v_n] .$$

Proof: The proof is straightforward, and the details are omitted. ∎

Theorem 6.6.8 *Let $\eta \in \eth\check{C}_{a;\check{D}}^+$. Then $\ell[a, \eta] = \eth\check{C}_{a;\check{D}}^+ \cap \eth\check{C}_{\eta;\check{D}}^-$.*

Proof: By definition, $\ell[a, \eta] \subset \eth\check{C}_{a;\check{D}}^+$. From part 4) of Properties 6.6.1.2, $\ell[a, \eta] \subset \eth\check{C}_{\eta;\check{D}}^-$. Therefore

$$\ell[a, \eta] \subset \eth\check{C}_{a;\check{D}}^+ \cap \eth\check{C}_{\eta;\check{D}}^- . \tag{6.24}$$

Let $\{w_n\}$, $w_{n+1} \ll w_n$, $w_n \in C_{a;D}^+$ be a sequence that converges to η. Then

$$\check{C}_{\eta;\check{D}}^- = \bigcap_{n \in \mathbb{N}} \check{C}_{w_n;\check{D}}^- ,$$

so that

$$\check{C}_{a;\check{D}}^+ \cap \check{C}_{\eta;\check{D}}^- = \bigcap_{n \in \mathbb{N}} \left(\check{C}_{a;\check{D}}^+ \cap \check{C}_{w_n;\check{D}}^- \right)$$
$$= \bigcap_{n \in \mathbb{N}} \check{I}[a, w_n]$$
$$= \ell[a, \eta]$$

where the last line follows from Proposition 6.6.5. Combining with the inclusion (6.24), we have the desired result. ∎

6.6.1.3 New Light Ray Segments on $\eth\check{C}_{\xi;\check{D}}^+$

Let $\xi \in \check{D}$ and $\eta \in \eth\check{C}_{\xi;\check{D}}^+$. Set

$$L = \eth\check{C}_{\xi;\check{D}}^+ \bigcap \eth\check{C}_{\eta;\check{D}}^- . \tag{6.25}$$

Then

Theorem 6.6.9 *L is totally ordered by the relation $<$ (equivalently, by the relation $>$).*

Proof: The proof depends on the fact that L is a closed, and therefore complete subset of $\check{M}$.

The relation $<$ defines a partial order on $\check{D}$, and therefore on L. Let K be a maximal totally ordered (by $<$) subset of L that has ξ as its smallest and η as its largest element, and has the property that if $z \in L \setminus K$ then there exists $x \in K$ such that x and z are spacelike to each other. Now:

1. If $K = L$ then there is nothing to prove.
2. Suppose, therefore, that $K \neq L$, and consider the completion $\widetilde{K}$ of K. Since L is complete and $K \subset L$, it follows that $K \subset \widetilde{K} \subset L$. Again, there are two possibilities:

$$\text{a) } \widetilde{K} \neq K.$$
$$\text{b) } \widetilde{K} = K.$$

a) Suppose that $\widetilde{K} \neq K$, and $y \in \widetilde{K} \setminus K$. Then x (the point spacelike to y) defines a section of K, i.e., a pair of subsets $A, B \subset K$ such that $A \cap B = \emptyset$, $A \cup B = K$ and $z \in A$, $z' \in B \Rightarrow z < z'$. Then there exist Cauchy sequences $\{z_i\}$ in A and $\{z'_j\}$ in B that converge to $y \in \widetilde{K}$. But then $\check{C}^+_{y;\check{D}} = \cap \check{C}^+_{z_i;\check{D}}$ and $\check{C}^-_{y;\check{D}} = \cap \check{C}^-_{z'_j;\check{D}}$, i.e., $y \in K$, a contradiction. This rules out the possibility $\widetilde{K} \neq K$. The proof is somewhat simpler if one of the sets $\{z_i\}$ and $\{z'_j\}$ is finite, for the finite set contains a maximal and minimal element.

b) If $\widetilde{K} = K \neq L$, then there must be gaps in K, i.e., pairs of points y, y' such that $y < y'$, $y \neq y'$ and $\partial \check{C}^+_{y;\check{D}} \cap \partial \check{C}^-_{y';\check{D}} = \emptyset$. But this is ruled out by Lemma 6.5.15, which means that the possibility $\widetilde{K} = K \neq L$ is also excluded. The only possibility that remains is $K = L$.

■

In view of Theorem 6.6.9, we define:

Definition 6.6.10 (Segments $\ell[\xi, \eta]$)

If $\xi, \eta \in \check{D}$, $\eta \in \partial \check{C}^+_{\xi;\check{D}}$, $\xi \neq \eta$, then

$$\ell[\xi, \eta] = \partial \check{C}^+_{\xi;\check{D}} \cap \partial \check{C}^-_{\eta;\check{D}} .$$

As before, the orders on $\ell[\xi, \eta]$ will be denoted by $<^l$ (or $^l>$) and $<^{ll}$. □

Theorem 6.6.11 Let $\{a_n\}$, $a_{n+1} \gg a_n$ and $\{b_n\}$, $b_{n+1} \ll b_n$ be Cauchy sequences in D that converge to ξ and η respectively in $\check{D}$. Then

$$\ell[\xi, \eta] = \bigcap_{n \in \mathbb{N}} \check{I}[a_n, b_n] .$$

Proof: Since $a_m \ll b_n \ \forall \ m, n \in \mathbb{N}$, the $I[a_n, b_n]$ are nonempty. Since $\check{\partial}\check{C}^+_{\xi;\check{D}} \cap \check{\partial}\check{C}^-_{\eta;\check{D}} \subset \cap \check{I}[a_n, b_n]$, it suffices to prove that $\cap \check{I}[a_n, b_n]$ contains no interior point of $\check{C}^+_{\xi;\check{D}}$ or of $\check{C}^-_{\eta;\check{D}}$. Let x be an interior point of $\check{C}^+_{\xi;\check{D}}$. Then $x \notin \check{C}^-_{\eta;\check{D}}$, and there exist $p, q \in \check{C}^+_{\xi;\check{D}}$ such that $x \in \check{I}(p, q) \subset \check{I}[p, q] \subset \tau \check{C}^+_{\xi;\check{D}}$. Then there exists N such that p and b_n are spacelike to each other for $n \geq N$. For suppose that there is no such N. Then $p \in \check{C}^-_{b_n;\check{D}}$ for all $n \in \mathbb{N}$, i.e., $p \in \check{C}^-_{\eta;\check{D}}$, a contradiction. However, if p and b_n are spacelike to each other for $n \geq N$, then $p \notin \check{I}[a_n, b_n]$ for $n \geq N$, i.e., $x \notin \cap \check{I}[a_n, b_n]$. This contradiction establishes the desired result. A similar proof establishes that $\cap \check{I}[a_n, b_n]$ contains no interior point of $\check{C}^-_{\eta;\check{D}}$. ∎

6.6.1.4 Properties of $\ell[\xi, \eta]$

The limit $\ell[\xi, \eta]$ has the following properties:

1. $\ell[\xi, \eta]$ is totally ordered, by Theorem 6.6.9.

2. $\ell[\xi, \eta]$ is closed, because it is the intersection of the sets $\check{\partial}\check{C}^+_{\xi;\check{D}}$ and $\check{\partial}\check{C}^-_{\eta;\check{D}}$ that are closed in $\check{D}$.

3. $\ell[\xi, \eta] \subset \check{\partial}\check{C}^+_{\xi;\check{D}}$, by definition.

4. $\ell[\xi, \eta] \subset \check{\partial}\check{C}^-_{\eta;\check{D}}$, by definition.

5. Let $p \in \tau\widetilde{C}^+_{\xi;\check{D}} \cap D$ such that $b_n \notin C^-_{p;D}$ for $n \in \mathbb{N}$, where the b_n are as defined in Theorem 6.6.11. Then $\check{\partial}\check{C}^+_{p;\check{D}} \cap \ell[\xi, \eta]$ is a unique point.

   Proof: Since p is spacelike to every b_n, there is a backward ray l^-_p (in M) from p that intersects every $\beta C^-_{b_n;D}$. (The argument is the same that was used in the first paragraph of the proof of Lemma 6.5.15.) Let $\{w_n\} = l^-_p \cap \beta C^-_{b_n;D}$. Then $w_{n+1} <^{ll} w_n$, and the sequence $\{w_n\}$ is bounded below. Therefore it has a limit, say ζ_1, in $\check{l}^-_p$, and $\zeta_1 \in \check{l}^-_p \cap \ell[\xi, \eta]$.

   Suppose now that there is another backward ray l'^-_p from p that intersects every $\beta C^-_{b_n;D}$. Then it intersects $\ell[\xi, \eta]$ at a point ζ_2. But since $\zeta_1 \in \check{l}^-_p$ and $\zeta_2 \in \check{l}'^-_p$, it follows that ζ_1 and ζ_2 are spacelike to each other, a contradiction. Therefore $l^-_p = l'^-_p$ and $\zeta_1 = \zeta_2$. ∎

6. Let $\zeta \in \tau\check{C}^+_{\xi;\check{D}}$ such that $b_n \notin C^-_{\zeta;\check{D}}$ for $n \in \mathbb{N}$. Then $\check{\partial}\check{C}^+_{\zeta;\check{D}} \cap \ell[\xi, \eta]$ is a unique point.

   Proof: Let $\{v_n\}$, $v_n \in \tau\check{C}^+_{\xi;\check{D}}$, $v_n \ll v_{n+1}$ be a sequence of points in D that converges to ζ. By property 5) just above, $\check{\partial}\check{C}^-_{v_n;\check{D}} \cap \ell[\xi, \eta]$ is a unique point, say ζ_n. The point in question is the limit of the sequence ζ_n. ∎

6.6.1.5 Extending Light Ray Segments

The light rays l are globally defined in M. The segment $\ell[\xi, \eta]$ is defined locally in $\check{D}$. One has to extend the definition of ℓ globally by means of overlapping covers. The key to this extension is provided by the following lemma:

Lemma 6.6.12 Let $y \in \check{\partial}\check{C}^+_{x;\check{D}}$, $y \neq x$. Then the set $\check{\partial}\check{C}^+_{x;\check{D}} \cap \check{\partial}\check{C}^+_{y;\check{D}}$ is totally ordered by $<$.

Proof: Let $z \in \check{\partial}\check{C}^+_{x;\check{D}} \cap \check{\partial}\check{C}^+_{y;\check{D}}$. Then the light ray segments $\ell[x, z], \ell[y, z]$ are defined and ordered by $<$, and $\ell[x, z] \supset \ell[y, z]$. The result follows. ∎

Combined with overlapping covers, Lemma 6.6.12 provides a ready procedure for extending segments of the new light rays ℓ. We omit the straightforward details.

This completes the definition of new light rays in $\check{M}$.

6.6.2 Verifications

We shall now verify that $\check{M}$ is indeed an ordered space in our sense of the term. For ease of reference the axioms, the definition of a D-set and the special assumptions are collected together (without change of numbering) in Appendix C.

6.6.2.1 The Order Axiom

It is easy to see that the family of light rays in $\check{M}$ satisfies the order Axiom 3.1.2. Parts a) and b) are among the standard properties of $\mathbb{R}$. Part c), which states that light rays do not have end-points, follows from the exclusion of boundary points in the definition of order completion. Part d) follows from the facts that order in $\check{M}$ is defined by extending the order relations $<^l, \ll$ and $<$ on M to $\check{M}$, and that these extensions never reverse the direction.

6.6.2.2 The Identification Axiom

If l and l' are two distinct light rays in M, then the identification axiom holds obviously for their completions $\check{l}$ and $\check{l}'$. Property 6.6.1.4, part 5 states that two light rays can intersect only once in a $\check{D}$-set. Finally, the identification axiom is used explicitly as a defining tool, to extend the definition of the segment $\ell[\xi, \eta]$ (see Sect. 6.6.1.5), and therefore it cannot be violated by pairs of new rays ℓ, ℓ'.

6.6.2.3 Definition of Forward and Backward Cones

The local cones $\check{C}^{\pm}_{x;\check{D}}$ were defined topologically (Def. 6.4.5) in Sect. 6.4. We begin by establishing their order-theoretic characterization.

Lemma 6.6.13 *Let* $y \in \check{C}^{+}_{x;\check{D}}$*. Then:*

1. *If* $y \in \check{\partial}\check{C}^{+}_{x;\check{D}}$ *then there exists a light ray that joins* x *and* y*.*
2. *If* $y \in \tau\check{C}^{+}_{x;\check{D}}$ *then there exist ascending l-polygons from* x *to* y*.*

The same assertions hold with order reversed.

Proof: From the definition of $\check{C}^{\pm}_{x;\check{D}}$, it follows that $y \in \check{C}^{+}_{x;\check{D}} \Leftrightarrow x \in \check{C}^{-}_{y;\check{D}}$.

1. Let $y \in \check{\partial}\check{C}^{+}_{x;\check{D}}$. If $x, y \in D$ then, by definition, $\lambda(x,y)$. If $x \in M$ but $y \notin M$ then, from Theorem 6.6.8, $\check{\partial}\check{C}^{+}_{x;\check{D}} \cap \check{\partial}\check{C}^{-}_{y;\check{D}} = \ell[x,y]$. If $x, y \notin M$ then, from Def. 6.25 and Theorem 6.6.9, $\check{\partial}\check{C}^{+}_{x;\check{D}} \cap \check{\partial}\check{C}^{-}_{y;\check{D}} = \ell[x,y]$.
2. We disregard the trivial case $x, y \in M$. Let $y \in \tau\check{C}^{+}_{x;\check{D}}$. Then any backward ray from y intersects $\check{\partial}\check{C}^{+}_{x;\check{D}}$ at exactly one point (Property 6.6.1.2(6) or Properties 6.6.1.4(5, 6), as the case may be). It now follows from 1) above that x is connected to y by an ascending l-polygon with two sides. Then x can be connected to y by ascending polygons with n sides for any $n \geq 2$.

The same proofs hold with order reversed. ∎

We now define cones globally as follows:

Definition 6.6.14

$$C^{+}_{x} \equiv \bigcup_{\substack{P \ni x \\ P \text{ ascending}}} P$$

$$C^{-}_{x} \equiv \bigcup_{\substack{P \ni x \\ P \text{ descending}}} P$$

and

$$C_{x} \equiv C^{+}_{x} \bigcup C^{-}_{x}.$$

□

6.6.2.4 The Cone Axiom

Earlier, we gave an example (Example 3.2.9) of a space – the anti-de Sitter space – which was locally $\mathbb{R}^2$, but did not satisfy the cone axiom. The same example could be constructed using $\mathbb{F}$, the field of real algebraic numbers (which is denumerable) instead of $\mathbb{R}$. This shows that the cone axiom is independent of completeness or incompleteness. In the example cited, the failure of the cone axiom can be traced to the global topology of the space.

However, it may be proved that the cone axiom holds in $\check{M}$ if it holds in M. The proof is as follows.

Let M be an ordered set which satisfies the cone axiom (i.e., $a \ll b \Rightarrow b \not\ll a$), $\check{M}$ its order completion and $x \in \check{M}$. Set

$$Q = \check{C}_x^+ \cap \check{C}_x^- \setminus \{x\} . \tag{6.26}$$

We shall assume that $Q \neq \emptyset$ and derive a contradiction.

Lemma 6.6.15 *If $Q \neq \emptyset$, then for any $a, b \in M$ such that $a \ll x \ll b$,*

$$\check{I}[a,b] \cap Q \neq \emptyset .$$

Proof: Let $y \in Q$. Since y belongs to $\check{C}_x^-$, there exists a descending l-polygon $P(y, \ldots, x)$ from y to x. Traversing P in the opposite direction we obtain an ascending l-polygon $P(x, \ldots, y)$. Since y also belongs to $\check{C}_x^+$, it follows that P also belongs to $\check{C}_x^+$. Hence $P \subset \check{C}_x^+ \cap \check{C}_x^-$, which proves the lemma. ∎

Theorem 6.6.16
$$Q = \emptyset .$$

Proof: Choose $z \in Q$ and let $\{a_n\}, \{b_n\}$ be sequences in M such that $a_n \to x, b_n \to x$, and
$$a_n \ll a_{n+1} \ll x \ll b_{n+1} \ll b_n .$$
Also, let $\{c_n\}, \{d_n\}$ be sequences in M such that $c_n \to z, d_n \to z$, and

$$c_n \ll c_{n+1} \ll z \ll d_{n+1} \ll d_n .$$

Then $I[a_n, b_n] \cap I[c_m, d_m] = \emptyset$ for sufficiently large n and m, and hence by Theorem 6.5.11, $\check{I}[a_n, b_n] \cap \check{I}[c_m, d_m] = \emptyset$ for such n, m. But $z \in kD_{x,\check{D}}^+ \subset \check{D}_{a_n,\check{D}}^+$ and $z \in \check{D}_{x,\check{D}}^- \subset \check{D}_{b_n,\check{D}}^-$ implies $z \in \check{I}[a_n, b_n]$ for all n, which is a contradiction. ∎

6.6.2.5 D-sets

We shall prove that the sets $\check{D} = \text{int}\,\widetilde{D} \subset \check{M}$, where D is a D-set in M, are indeed D-sets in $\check{M}$, i.e., they satisfy the conditions a)–f) of Definition 4.2.1 (also reproduced in Appendix C).

a) $x, y \in \check{D}, x \ll y \Rightarrow \check{I}[x,y] \subset \check{D}$.
   Let U_x, U_y be open sets in $\check{M}$ such that $x \in U_x \subset \check{D}$, $y \in U_y \subset \check{D}$ and $U_x \cap U_y = \emptyset$. There exist Cauchy sequences $\{a_n\} \subset U_x$ and $\{b_n\} \subset U_y$ that converge to x and y respectively such that $a_{n+1} \gg a_n$ and $b_{n+1} \ll n_n$ for all $n \in \mathbb{N}$. Then

$$\check{I}[x,y] = \left(\cap \check{C}^+_{a_m;\check{D}}\right) \cap \left(\cap \check{C}^-_{b_n;\check{D}}\right)$$

$$= \bigcap_{n\in\mathbb{N}} \left(\check{C}^+_{a_n;\check{D}} \cap \check{C}^-_{b_n;\check{D}}\right)$$

$$= \bigcap_{n\in\mathbb{N}} \check{I}[a_n,b_n] .$$

From the separation Lemma 6.5.8, $\check{I}[a_1,b_1] \subset \check{I}(a_0,b_0)$. Since $\check{I}(a_0,b_0) \subset \check{D}$, the result follows.

b) Since $\check{D}$ is an open set, the intersection of a light ray (either $\check{l}$ or ℓ) cannot have a largest or a smallest element (with respect to order on the ray).

c) Let $x,y,z \in \check{D}$, $x \ll y \ll z$. Then $\check{l}^+_y$ (or ℓ^+_y, as the case may be) intersects $\partial \check{C}^-_{z;\check{D}} \setminus \{z\}$ at a single point and $\check{l}^-_y$ (or ℓ^-_y, as the case may be) intersects $\partial \check{C}^+_{x;\check{D}} \setminus \{x\}$ at a single point; these are precisely the properties 5 and 6 of $\ell[a,\eta]$ and $\ell[\xi,\eta]$ that were established earlier (Sects. 6.6.1.2 and 6.6.1.4).

d) The convexity axiom: If two distinct points lie on a ray $\check{l}$, this ray will clearly satisfy the convexity axiom with respect to $\partial \check{C}^\pm_{x;\check{D}}$ for any $x \in \check{l} \cap \check{D}$. New rays ℓ are defined segmentwise by intersections $\ell[x,y] = \partial \check{C}^+_{x;\check{D}} \cap \partial \check{C}^-_{y;\check{D}}$, and therefore the ray ℓ will also satisfy the convexity axiom with respect to $\partial \check{C}^\pm_{z;\check{D}}$ for any $z \in \ell \cap \check{D}$.

e) By construction, if $y \in \partial \check{C}^+_{x;\check{D}}$ or $x \in \partial \check{C}^-_{y;\check{D}}$, there exists only one light ray that passes through x and y.

f) The cardinality of the set of rays through $x \in \check{M}$ cannot be smaller than the cardinality of the set of rays through $a \in M$.

6.6.2.6 The Local Structure Axiom

It was established above that every $\check{D}$-interval is a D-set. By definition, every point ξ introduced by the completion is contained in a nonempty open $\check{D}$-interval $\check{I}(a,b)$. Observe that the family of $\check{D}$-intervals $\{\check{I}(a,b)|a,b \in D \subset M, a \ll b\}$ is a base for the order topology of $\check{M}$.

This concludes our verification of the axioms in $\check{M}$. We have, in the process, justified the notations and terminology that had been carried over from M to $\check{M}$.

6.7 Remarks on the Assumptions 6.4.1 and 6.5.1

1. In Sect. 6.4, we made the assumption of local precompactness (6.4.1) to ensure that the order uniformity on a D-set was the only uniformity on D compatible with its topology. However, we made no explicit use of this

assumption for extending the order structure to the order completion $\check{M}$ of M. What we did use extensively was the local *topological* structure of $\check{M}$, which should be common to all uniform completions of M. It is therefore possible that the results of this chapter hold largely, if not in their entirety, without the assumption of local precompactness.

2. In Sect. 6.5, we made the Assumption 6.5.1 that the topology of M satisfies the first axiom of countability. This enabled us to use sequences in the constructions and the proofs. It should, however, be noted that wherever sequences have been used, they may be replaced by nets with little more than notational changes. It is therefore likely that the results of this chapter hold without the assumption of first countability.

We have not studied either of these questions.

7

Spaces with Complete Light Rays

In this chapter we shall establish some results that hold in ordered spaces in which light rays are complete (in which case they are locally homeomorphic with $\mathbb{R}$),[1] but the space as a whole need not be order complete. We begin with an example of such a space, which is infinite-dimensional. We should add that we know of no example of a finite-dimensional ordered space in which light rays are complete but the space itself in not (order) complete. The phenomenon could be peculiar to infinite-dimensional spaces.

The properties established in this chapter are the connectivity properties (Sect. 7.2) and a miscellany of global properties (Sect. 7.3).

In the sequel, there will be no further need to adhere to the special notations of the previous chapter (Sect. 6.4), particularly Def. 6.4.1. We shall therefore revert to our original notations, which will apply henceforth to all ordered spaces, be they order complete or otherwise.

The space of the Example 7.1 given below is second countable, and therefore first countable. However, the results that follow are established without the first countability Assumption 6.5.1.

7.1 An Infinite-dimensional Space

In the following example, the light rays are complete, but the space itself fails to be order complete.

Example 7.1.1 Consider the space $\mathfrak{S}$ of real sequences $s = \{x_n\}$, $n = 0, 1 \ldots$ in which only a finite number of entries are nonzero, and define on $\mathfrak{S}$ the

---

[1] Since an ordered space is Tychonoff and the Tychonoff property is hereditary [57], every light ray is Tychonoff in the subspace topology, and can therefore be completed (uniformly) independently of the rest of the space. If the light ray has an overlapping D-cover, then its completion will be locally homeomorphic with $\mathbb{R}$. One can therefore imagine an ordered space in which the light rays are complete, but the space itself is not order complete.

H.-J. Borchers and R.N. Sen: *Mathematical Implications of Einstein–Weyl Causality,*
Lect. Notes Phys. **709**, 95–101 (2006)
DOI 10.1007/3-540-37681-X_7

Minkowski form

$$|x|^2 = -x_0^2 + \sum_{k \in \mathbb{N}^+} x_k^2 \ .$$

If z is any point such that $|z|^2 = 0$, then the set of points $\{\lambda z\}, \lambda \in \mathbb{R}$ defines a light ray. It is clearly locally homeomorphic with $\mathbb{R}$. However, the sequence of points $\{s_N\}_{N \in \mathbb{N}^+}$, where

$$s_N = \begin{cases} \dfrac{1}{n^2}, 1 \le n \le N \\ \\ 0, \quad n > N, \end{cases}$$

which is clearly Cauchy, does not converge in $\mathfrak{S}$. Therefore $\mathfrak{S}$ is not complete. Its completion is the space l_2 of all convergent sequences, which is a Hilbert space without holes or boundaries. Therefore the order completion of $\mathfrak{S}$ is the same as its completion, and therefore $\mathfrak{S}$ is not order complete.

7.2 Connectivity Properties

This fact that light rays are locally homeomorphic with $\mathbb{R}$ leads immediately to the following connectivity results:

Theorem 7.2.1 *Let M be an ordered space in which light rays are complete. Then M is path-connected.*

Proof: If every light ray is complete, then every l-polygon becomes a path, and the l-connectedness condition becomes the path connectedness condition. ∎

Theorem 7.2.2 *Let M be an ordered space in which light rays are complete. Then M is connected.*

Proof: For any separation of M into disjoint open subsets will break up light rays into disjoint open segments – which is impossible. ∎

7.3 Some Global Results

The results that will be established below are global. They may be classified as follows: 1) Local properties that extend globally, and 2) global properties that have no local counterparts.

7.3.1 Local Properties that Extend Globally

We begin with the following proposition:

Proposition 7.3.1

Let M be an ordered space in which light rays are complete. Let $x, y \in M$ and $x \neq y$. Then there exists an l-polygon $P(x_0, x_1, \ldots, x_n)$, $n \in \mathbb{N}$, where $x_0 = x$ and $y = x_n$, such that any two of its successive vertices x_i, x_{i+1}, $i = 0, 1, \ldots, n-1$, lie in a D-set.

Proof: It suffices to prove that every closed light ray segment lies in a D-set.

Since the light rays in M are complete, closed segments of light rays l, being homeomorphic to closed intervals in $\mathbb{R}$, are compact in the order topology $<^l$ on l. So therefore is the l-polygon $P(x_0, \ldots, x_n)$. Since the order topology on l is equivalent to the subspace topology inherited from M, it follows that every D-cover of P has a finite subcover.

Suppose that the segment $l[x_n, x_{n+1}]$ is covered by m D-sets D_i, $i = 1, \ldots, m$. Consider the two D-sets D_i, D_{i+1} and choose points y_i, $i = 1, 2, 3$ in $l[x_n, x_{n+1}]$ with $y_1 \in D_{i-1} \cap D_i$, $y_2 \in D_i \cap D_{i+1}$ and $y_3 \in D_{i+1} \cap D_{i+2}$. Then we can find points $p, q \in D_i$ such that $p \ll y_1, q \gg y_2$ and $r, s \in D_{i+1}$ with $r \ll y_2, s \gg y_3$. This implies that $l[y_1, y_2] \subset I(p, q)$ and $l[y_2, y_3] \subset I(r, s)$. Let $p_0 = \beta C_p^+ \cap l_{y_1, y_2}$ and $s_0 = \beta C_s^- \cap l_{y_2, y_3}$. Then we can find $a \in l(p, p_0)$ and $b \in l(s_0, s)$ such that $I(a, b) \subset I(p, q) \cup I(r, s)$. For arbitrarily chosen points a, b it might happen that a light ray $l^{(2)}$ other than $l_{x_i, x_{i+1}}$ intersects $l_{x_i, x_{i+1}}$ twice. Call these points of intersection z_1, z_2. Since we cannot exclude the possibility that the two light rays $l^{(2)}$ and $l_{x_i, x_{i+1}}$ are close to each other, it might happen that $l^{(2)}[z_1, z_2] \subset I(a, b)$. However, the two light rays are not arbitrarily close to each other, and therefore we can find points a_1, b_1 with $a <^l a_1 <^l p_0$ and $s_0 <^l b_1 <^l b$ such that no two light rays intersect twice in $I(a_1, b_1)$. The other properties of D-sets (Def. 4.2.1) are easily verified.

If $l[x_k, x_{k+1}]$ is covered by n D-sets, $n > 2$, then one has to repeat the construction $n - 1$ times to obtain a single open order interval that covers $l[x_k, x_{k+1}]$.   ∎

In Chap. 3 we gave an example (Fig. 3.6 and Remarks 3.2.23) of an ordered space that did not have the property S (Def. 3.2.21; if $x, y \in M$ then $y \in \tau C_x^- \Leftrightarrow x \in \tau C_y^-$). In the example, failure of the property S could be traced immediately to the fact that M consisted of two (open) half-planes joined by a single point, the origin. However, this example does not satisfy the local structure axiom; there is no D-set that contains the origin. This suggests that if the two half-planes are joined, not by a single point, but by a D-set, no matter how "small", property S will be recovered.

Theorem 7.3.2 *Let M be an ordered space in which light rays are complete. Then M is an S-space, that is, if $p, q \in M$ then $q \in \tau C_p^+ \Leftrightarrow p \in \tau C_q^-$.*

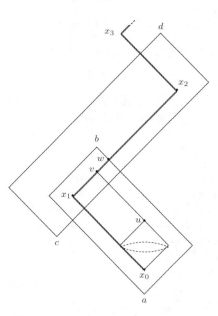

Fig. 7.1. Establishing Property S

Proof: Since every light ray is locally homeomorphic with $\mathbb{R}$ and since every point in M is contained in a D-set, it follows that every light ray has an overlapping cover. Suppose that $p \in \tau C_q^-$. Then there exist descending l-polygons from q to p which are also ascending l-polygons from p to q. In view of Proposition 7.3.1, we may, without loss of generality, confine our attention to those with sides that are contained in D-sets. Let then $P(p = x_0, x_1, \ldots, x_n = q)$ be an ascending l-polygon, with $x_i <^l x_{i+1}$, $i = 0, \ldots, n-1$, $\sim\!\lambda(x_i, x_{i+2})$, $i = 0, \ldots, n-2$ such that $l[x_i, x_{i+1}] \subset D_i$, where D_i is a D-set.

Consider the segments $l[x_0, x_1]$ and $l[x_1, x_2]$. We may find points a, b, c, d such that $a \ll b$, $c \ll d$ and

$$l[x_0, x_1] \subset I(a,b) \subset I[a,b] \subset D_1 \,,$$
$$l[x_1, x_2] \subset I(c,d) \subset I[c,d] \subset D_2$$

(see Fig. 7.1). Since $x_1 \in I(a,b)$, the segment $l[x_1, x_2]$ intersects $\beta I[a,b]$ at a single point, say w. Then $w \gg x_0$, and therefore there exists u such that $w \gg u \gg x_0$. Then $\beta C_u^+ \cap l[x, w]$ is a single point, say v, and there is, trivially, an ascending l-polygon $P(u, v, x_2, \ldots, x_n)$. It follows that, for *any* point $\omega \in S(x, u)$, there is an ascending l-polygon $P(x_0, \omega, u, v, x_2, \ldots, x_n)$, i.e., $q \in \tau C_p^+$.

The reverse implication is proved in the same manner.   ∎

Finally, we establish that Theorem 4.5.4 holds globally, and not just in D-sets:

Theorem 7.3.3 *Let M be an ordered space in which light rays are complete, and $x \in M$. Then $\tau C_x^+ = \operatorname{int} C_x^+$, and the same with order reversed.*

Proof: It will suffice to prove the assertion for positive cones.

Let $y \in \tau C_x^+$. Then there exists an l-polygon $P(x, x_1, \ldots, x_n, y)$ such that $l[x_i, x_{i+1}] \subset D_i$, where D_i is a D-set, $i = 1, 2, \ldots n$. Consider the segment $l[x_n, y] \subset D_n$. By assumption, $y \in \tau C_x^+$, but for x_n there are the two possibilities, $x_n \in \tau C_x^+$ or $x \in \beta C_x^+$. Choose a point $p \in l(x_n, y)$. Then (in either case) $p \in \tau C_x^+$. Next, choose a light ray $l_p \neq l_{x_n, y}$. Then l_p^- intersects C_x^+ in a nonempty closed segment, and βC_x^+ at a single point (which need not be named). Therefore there exists $q <^{ll} p$ such that $l[q, p] \subset \tau C_x^+$. We may, without loss of generality, choose q close enough to p so that $q \in \tau C_x^+ \cap D_n$. This implies that $q \ll y$. From the definition of the cone, there exists $r \in \tau C_x^+ \cap U_n$ such that $r \gg y$. Then $y \in I(q, r) \subset \tau C_x^+$, which proves that τC_x^+ is open. ∎

It should be noted that the statement $\beta C_x^\pm = \partial C_x^\pm$ will *not* be true globally; see the example in 3.2.16.

Theorem 7.3.4 *Let M be an ordered space in which light rays are complete. Let $x, y, z \in M$ such that $y, z \in \beta C^+ x$ and $x <^l y, y <^l z$. Then $z \in l_{x,y}$.*

Proof: Let U be a D-set containing y, and $p \in l[x, y] \cap U, p \neq y$. Next, choose $q \in l[y, z] \cap U, q \neq y$. Since $\tau C_p^+ \subset \tau C_x^+$ and $q \in \beta C_x^+$ it follows that $q \in \beta C_p^+$. Therefore, by the convexity axiom, $l_{p,y} = l_{y,q}$. This implies that $l_{p,y} = l_{y,q} = l_{x,z}$. ∎

Corollary 7.3.5 *Let M be an ordered space in which light rays are complete, and $y \in \beta C_x^+ \subset M$. Then there exists a light ray l_x through x that passes through y.*

Proof: Since $y \in \beta C_x^+ \subset C_x^+$, there exist ascending l-polygons from x to y. Assume that $x_i \in \beta C_x^+$. Then $x_{i+1} \in \beta C_x^+$, because $x_{i+1} \in \tau C_x^+$ would imply $y \in \tau C_x^+$, contradicting the assumption. It then follows from Theorem 7.3.4 that $x, x_i, x_{i+1}, \ldots, y$ lie on the same light ray. ∎

The method of proving Theorem 4.2.12, p. 37 allows one to generalize Lemma 4.2.18, p. 40 as follows:

Corollary 7.3.6 *Let $y \in \beta C_x^+ \subset M$. Let l_y be a light ray through y such that $l_y \neq l_{x,y}$. Then $l_y^+ \setminus \{y\} \subset \tau C_x^+$.* ∎

7.3.2 Global Properties Without Local Counterparts

If we look at a light ray globally, and not just in a D-set, we find that we have not excluded the possibility that a light ray through x starts on βC_x^+ and at some later time (outside a D-set) enters into the interior of C_x^+. The fact that this can happen can be seen from the example of the cylinder based on the circle S_1. Here the light rays are spirals that make an angle of $\pi/4$ with the

horizontal circles, and two light rays through x meet infinitely many times.[2]
The next result shows that this is a general feature (which holds even if the
ordered space is not complete):

Lemma 7.3.7 *Let M be an ordered space in which light rays are complete,
and such that there are more than two light rays through each point. Let $x, y \in
M$, $x <^l y$. If two or more light rays through x intersect at y, then*

$$l_y^+ \setminus \{y\} \subset \tau C_x^+$$

for each of these light rays.

Proof: Assume, to the contrary, that there exists a point $z \in l_y^+$ that belongs
to βC_x^+. Let l' be another light ray connecting x with y. Then $l'[x, y], l[y, z]$ is
an l-polygon connecting x with z. Since $z \in \beta C_x^+$, Theorem 7.3.5 asserts that
$l = l'$, a contradiction.   ∎

We shall now show that the situation described in Lemma 7.3.7 is the only
one in which a light ray can "dive" into the interior of a light cone.

Theorem 7.3.8 *Let M be as in Lemma 7.3.7.[3] Let $x \in M$ and $l_x^{(1)}$ a light
ray through x. Assume that there exists a point $z \in l_x^{(1)+}$ such that $x \ll z$ and
$z \in \tau C_x^+$. Then there exists $y \in l_x^{(1)+}$, $y <^l z$, $y \neq z$ such that x and y are
connected by at least two different light ray segments.*

Proof: First, consider the ray $l_x^{(1)}$ as a topological subspace of M. The intersection $l_x^{(1)} \cap \tau C_x^+$ is open in M, because τC_x^+ is open in M. It is nonempty
because, by hypothesis, it contains the point z. Therefore the set

$$\Lambda = l_x^{(1)+} \setminus \{l_x^{(1)+} \cap \tau C_x^+\}$$

is closed in $l_x^{(1)}$. Since Λ is totally ordered, it contains a maximal element y,
and every $t \in l_x^{(1)}$ with $y <^l t$, $t \neq y$ is contained in τC_x^+.

Observe that $C_y^+ \setminus \{y\} \subset \tau C_x^+$. Suppose that this is not true. Then there
exist points $p \in \beta C_x^+ \cap C_y^+ \setminus \{y\}$. However, if $l[y, p], l[x, y] \subset \beta C_x^+$, then
$p \in l_{x,y}^{(1)}$, i.e., $C_y^+ \setminus \{y\} \subset l_x^{(1)}$, which is impossible.

Let now $\mathcal{V}$ be a D-cover for $l_x^{(1)}[x, z]$. Set $W = \cup_{V \in \mathcal{V}} V$ and choose $q \in
\{\partial \{\beta C_x^+ \cap W\}\} \cap W$ with $q \notin l_x^{(1)}$. There are three possibilities:

---

[2] The formation of images by ideal lenses, i.e., lenses that have no aberrations,
would provide physical examples; all forward rays from the object point meet
again at the image point.

[3] In two-dimensional examples such as $S_1 \times \mathbb{R}$ considered above, a suitably-placed
hole in M would obstruct the second light ray through x from intersecting the
first.

i) $q <^l y$.

ii) $q \ll y$.

iii) q and y are spacelike.

In case i) the points (y, q, x) lie on the same light ray, which is the ray $l^{(2)}$ we are looking for. Case ii) is impossible, since it would imply that $y \in \tau C_x^+$, contradicting the construction of y. In case iii) we choose $s \notin l_{x,q}$ with $s > q$, which implies $s \gg x$. As usual we set $S(x, s) = \beta C_x^+ \cap \beta C_s^-$, and again find that there are three possibilities:

a) There exists $u_0 \in S(x, s)$ with $u_0 <^l y$ and $u_0 \notin l^{(1)}$.

b) $u \in S(x, s)$ and $u \ll y$ and $u \ll y$, with the possible exception of one point u_0. The exception occurs if $u_0 \in l^{(1)}$.

c) $u \in S(x, s)$, u and y are spacelike, with the possible exception of one point u_0.

In case a) one has $l_{x,u_0} = l_{u_0,y} = l^{(2)}$. In case b), $u \ll y$ implies that $l_{x,u} \cap \beta C_y^- = p$ is a single point. This in turn implies that $l_{x,p} = l_{p,y} = l^{(2)}$. In case c) we choose $t_n \in l^{(1)}$, $y <^l t_n$ such that $t_{n+1} <^l t_n$ and $\{t_n\}$ converges to y. This construction implies that $t_n \gg x$. Therefore the intersection $l_{x,u} \cap \beta C_{t_n}^- = p_n$ consists of a single point. Since t_n is monotonic decreasing and is bounded below by y we conclude that $p_n \in l_{x,u}$ is also monotonic decreasing and bounded below. Hence this sequence converges to a point $p <^l y$. Then, as before, $l_{x,p} = l_{p,y} = l^{(2)}$. ∎

We end this chapter with the remark that, as shown in [11], one can construct timelike curves connecting any two points $a, b \in U \subset M$, $a \ll b$, where U is a D-set and M an ordered space in which light rays are complete. The timelike curves so constructed turn out to be locally homeomorphic with $\mathbb{R}$. For details, the reader is referred to the original article.

Consequences of Order Completeness

In this chapter we shall investigate some consequences of order completeness. We remind the reader that there will be no further need to adhere to the special notations of Sect. 6.4, and we shall therefore revert to our original notations, which will apply henceforth to all order complete spaces.

The key to the results of this chapter is the existence of timelike curves. Section 8.1 gives a constructive proof of this assertion. Timelike curves are then used to establish other properties of order complete spaces, leading to the main result that open D-intervals in locally compact order complete spaces have a differentiable structure. This is achieved in Sect. 8.4.

The first countability Assumption 6.5.1 is not invoked in the present chapter, but the local precompactness Assumption 6.4.1 is used quite explicitly in Sect. 8.3.

8.1 Timelike Curves

We begin with the definition of *timelike curves*.

Definition 8.1.1 *Let U be a D-set, $x, y \in U$, $x \ll y$. A continuous curve $\Theta : [0,1] \to U$ such that $\Theta(0) = x$, $\Theta(1) = y$ is called a timelike curve iff $t_1, t_2 \in [0,1]$, $t_1 < t_2$ implies that $\Theta(t_1) \ll \Theta(t_2)$.* $\square$

In this section we shall establish the existence of timelike curves. The standard maps ρ and σ which were defined in 5.1.2 will play a central role in this process. Examination of the definition (Def. 5.1.2) and the fundamental properties (Propositions 5.1.3 and 5.1.4) of the maps ρ and σ shows that they carry over unchanged to order complete spaces. As the proof of existence of timelike curves is purely constructive, we shall present it as a construction rather than as a theorem.

The construction is a simplified version of the one given in [11], the simplification arising from the fact that we are dealing with order complete spaces,

H.-J. Borchers and R.N. Sen: *Mathematical Implications of Einstein–Weyl Causality*,
Lect. Notes Phys. **709**, 103–127 (2006)
DOI 10.1007/3-540-37681-X_8

which was not assumed in [11].[1] The accompanying diagrams (Figs. 8.1, this page and 8.2, page 106) have been modified accordingly.

Let U be a D-set, $a, b \in U$ and $a \ll b$. Choose two different points $p, q \in S(a, b)$, and consider the closed light ray sections $l[a, p]$ and $l[a, q]$. Since $l[a, p]$ and $l[a, q]$ are homeomorphic to the interval $[0, 1] \subset \mathbb{R}$, we may parametrize them as follows.

Let x, y be the generic points on the segments $l[a, p]$ and $l[a, q]$ respectively. We assign to x the coordinates $(\alpha, 0)$ and to y the coordinates $(0, \beta)$, where $\alpha, \beta \in [0, 1] \subset \mathbb{R}$. Then $a = (0, 0)$, $p = (1, 0)$ and $q = (0, 1)$. We set $b = (1, 1)$ (Fig. 8.1), and proceed as follows:

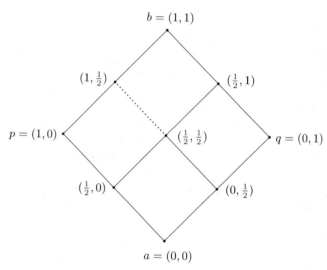

Fig. 8.1. Construction of timelike curves: Step 1

1. a) Define the point $(\frac{1}{2}, 1) \in l[q, b]$ by

$$(\tfrac{1}{2}, 1) = \beta C^+_{(1/2, 0)} \cap l[q, b]$$

and draw the light ray segment $l[(\frac{1}{2}, 0), (\frac{1}{2}, 1)]$.

---

[1] The construction of [11] was based on the assumption that light rays were complete, i.e., locally homeomorphic with $\mathbb{R}$; no assumption was made regarding the completeness or order completeness of M itself. As a result, considerable effort had to be invested to show that the "b-irrational points" (terminology of [11]) on timelike curves did, in fact, belong to M. This problem does not arise if M is assumed to be order complete.

b) Define the points

$$(\tfrac{1}{2}, \tfrac{1}{2}) = \beta C^+_{(0,1/2)} \cap l[(\tfrac{1}{2}, 0), (\tfrac{1}{2}, 1)]$$

and

$$(1, \tfrac{1}{2}) = \beta C^+_{(1/2,1/2)} \cap l[p, b]$$

and draw the segments $l[0, 1/2]$ and $l[(1/2, 1/2), (1, 1/2)]$ (Fig. 8.1). In the figure, the segment $l[(1/2, 1/2), (1, 1/2)]$ has been dotted, to indicate that this segment and the segment $l[(0, 1/2), (1/2, 1/2)]$ may belong to different light rays; due to the cushion problem,[2] the two light rays $l_{(0,1/2),(1/2,1/2)}$ and $l_{(1/2,1/2),(1,1/2)}$ need not coincide.

It follows immediately from the construction that

$$(0, 0) \ll (\tfrac{1}{2}, \tfrac{1}{2}) \ll (1, 1) .$$

c) Finally, use the standard maps

$$\rho : l[(\tfrac{1}{2}, 0), (1, 0)] \to l[(\tfrac{1}{2}, \tfrac{1}{2}), (1, \tfrac{1}{2})]$$

and

$$\rho : l[(0, \tfrac{1}{2}), (0, 1)] \to l[(\tfrac{1}{2}, \tfrac{1}{2}), (\tfrac{1}{2}, 1)]$$

to define numerical coordinates on the ranges $l[(1/2, 1/2), (1, 1/2)]$ and $l[(1/2, 1/2), (1/2, 1)]$ according to the rules $\rho(\alpha) = \alpha$ and $\rho(\beta) = \beta$ respectively.

2. We now have a figure consisting of four light ray quadrangles (Fig. 8.1), but only two of them will play a role in the next step. These are the quadrangles defined by the vertices

Lower quadrangle : $(0, 0)$, $(0, \tfrac{1}{2})$, $(\tfrac{1}{2}, \tfrac{1}{2})$, $(\tfrac{1}{2}, 0)$;

Upper quadrangle : $(\tfrac{1}{2}, \tfrac{1}{2})$, $(\tfrac{1}{2}, 1)$, $(1, 1)$, $(1, \tfrac{1}{2})$.

Notice that the light ray segments forming the lower sides of both these quadrangles have numerical coordinates defined on them. Therefore the procedure of the first step can be repeated in both of them. What one obtains by repeating the procedure is shown in Fig. 8.2. In the figure we have, for clarity, deleted the light ray segments and coordinates of the points from step 1 that played no role in step 2. We are then left with a figure with four light ray quadrangles along the "diagonal" connecting a with b, with the corresponding vertices ordered as follows:

$$(0, 0) \ll (\tfrac{1}{4}, \tfrac{1}{4}) \ll (\tfrac{1}{2}, \tfrac{1}{2}) \ll (\tfrac{3}{4}, \tfrac{3}{4}) \ll (1, 1) .$$

---

[2] The *cushion problem* is as follows: In an arbitrary order complete space, the ray that joins the points $(0, 1/2) \in l[a, q]$ and $(1, 1/2) \in l[p, b]$ need not intersect the ray segment $l[(1/2, 0), (1/2, 1)]$; this problem will be discussed in Chap. 9.

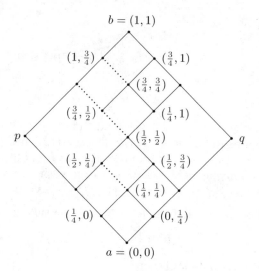

Fig. 8.2. Construction of timelike curves: Step 2

By iterating this procedure, we obtain a countable set of points

$$\Theta_0 = \left\{ \left(\frac{2k+1}{2m}, \frac{2k+1}{2m} \right) \middle| \, k = 0, \ldots, 2^{m-1} + 1, \, m \in \mathbb{N}^+ \right\}$$

which is linearly ordered by $\ll$. Set

$$x_\alpha = \left(\frac{2k+1}{2m}, \frac{2k+1}{2m} \right),$$

where α is a double index; $\alpha = (k, m)$. Then, if α, γ are such that $x_\alpha, x_\gamma \in \Theta_0$ and $x_\alpha \ll x_\gamma$, these exists $x_\beta \in \Theta_0$ such that $x_\alpha \ll x_\beta \ll x_\gamma$. The completion of Θ_0 is the sought-for timelike curve $\Theta[a, b] \subset M$ connecting a with b. It remains to show that there are no gaps in this curve.

To do this, let k, m be fixed. Although the point

$$\theta = \left(\frac{2k+1}{2m}, \frac{2k+1}{2m} \right) \in \Theta_0$$

is not connected to the points

$$\left(\frac{2k+1}{2m}, 0 \right) \in l[a, p], \quad \left(0, \frac{2k+1}{2m} \right) \in l[a, q]$$

by light rays, they are so connected, *by construction*, by descending *l*-polygons. Figure 8.3 shows these polygons explicitly for the points $\theta = (1/4, 1/4)$, $(1/2, 1/2)$ and $(3/4, 3/4)$. (In the figure, the vertices have been re-labelled for simplicity.) For example, the points p_1 and w_2 are connected by

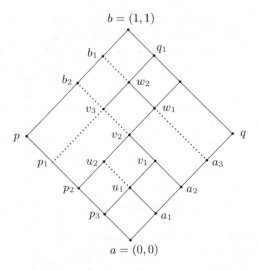

Fig. 8.3. Timelike curves: Reading off the maps Υ_x and Υ_y

the ascending l-polygon $P(p_1, v_3, w_2)$, the points a_3 and w_2 by the ascending l-polygon $P(a_3, w_1, w_2)$. Again, by construction, there exist order-preserving homeomorphisms

$$\Upsilon_x : l\left[a, \left(0, \frac{2k+1}{2^m}\right)\right] \to P\left[\left(0, \frac{2k+1}{2^m}\right), \dots, \left(\frac{2k+1}{2^m}, \frac{2k+1}{2^m}\right)\right],$$

$$\Upsilon_y : l\left[a, \left(\frac{2k+1}{2^m}, 0\right)\right] \to P\left[\left(\frac{2k+1}{2^m}, 0\right), \dots, \left(\frac{2k+1}{2^m}, \frac{2k+1}{2^m}\right)\right].$$
(8.1)

These homeomorphisms are glued together from pieces that are composite maps $\rho \circ \rho \circ \cdots \circ \rho$, and are cumbersome to write down explicitly. (The special cases

$$\Upsilon_x : l[a, a_3] \to P(p_1, v_3, w_2),$$

$$\Upsilon_y : l[a, p_1] \to P(a_3, w_1, w_2)$$

of the maps (8.1) may be read off Fig. 8.3 in a straightforward manner.) The fact that the maps Υ_x, Υ_y are homeomorphisms establishes that there are no gaps in $\Theta[a, b]$, for a gap in $\Theta[a, b]$ would imply a gap in $l[a, p]$, or a gap in $l[a, q]$, or both, which would contradict the known structure of closed light ray segments.

8.1.1 Topology of $\Theta[a, b]$

Since the curve $\Theta[a, b]$ is continuous in the geometrical sense, the following result holds:

Theorem 8.1.2 (Local structure of $\Theta(t)$) *The topology of the order $\ll$ on a timelike curve $\Theta[a, b]$ as defined above is the same as the subspace topology it inherits from M; Θ is locally homeomorphic with $\mathbb{R}$.*

Proof: By construction, there is an order-preserving bijection between the set $\{\Theta(t)|0 \leq t \leq 1\}$ furnished with the order $\ll$, and the closed real interval $[0, 1]$. An open order interval contained in $I(a, b)$ is either disjoint from the continuous curve Θ, or else intersects it on an open arc. The subspace topology on Θ is therefore the same as the topology generated by the order $\ll$ on it. ∎

8.2 Parametrization of D-Intervals

The aim of this and the following section is to establish that the interiors of D-intervals in locally compact ordered spaces are homeomorphic with the interiors of double cones in some finite dimensional Minkowski space. The proof requires several steps. In this section we shall establish those results that do not make explicit use of local compactness. We begin with a preliminary lemma:

Lemma 8.2.1 *Let M be an order complete space, $U \subset M$ a D-set, $a, b \in U$, $a \ll b$ and $\Theta[a, b]$ a timelike curve connecting a and b. Then, for any point $w \in I(a, b)$, there exist points $p, q \in \Theta(a, b)$, $p \ll q$ such that[3]*

$$\beta C_w^- \cap \Theta(a, b) = \{p\} \,,$$

$$\beta C_w^+ \cap \Theta(a, b) = \{q\}$$

so that

$$w \in S(p, q) \,.$$

Proof: Since, by Theorem 8.1.2, the subspace topology on Θ is the same as the order topology determined by the order $\ll$, the intersection of $\Theta[a, b]$ with any closed order interval properly contained in $I[a, b]$ is either empty or a closed segment of $\Theta[a, b]$. For any $w \in I(a, b)$, the order interval $I[w, b]$ has nonempty interior and is properly contained in $I[a, b]$. Therefore there exists a point $q \neq b$ such that $\Theta[a, b] \cap I[w, b] = \Theta[q, b]$. The point q cannot be an interior point of $I[w, b]$; if it were, then one would have $I(w, q) \cap \Theta[a, b] = \emptyset$, contradicting the fact that $\Theta[a, b]$ is a continuous curve. Therefore $q \in \beta I[w, b]$. Next, $q \notin \beta C_b^-$, for $q \in \beta C_b^-$ would imply that $\lambda(q, b)$, contradicting the fact that $q \in \Theta(a, b)$. Hence $q \in \beta C_w^+$.

The existence of the point p is established similarly. It follows immediately that $w \in S(p, q)$. ∎

---

[3] $\Theta(a, b)$ denotes the curve $\Theta[a, b]$, minus the end-points.

8.2.1 Boundaries of D-intervals. The Jordan-Brouwer Separation Property

The sphere $\mathbb{S}^{n-1} \subset \mathbb{R}^n$ separates $\mathbb{R}^n$ into an inner and an outer part that have $\mathbb{S}^{n-1}$ as their common boundary, and every simple curve that joins a point in the inner part with a point in the outer part intersects this common boundary. For $n = 2$, this is the classical Jordan curve theorem; for $n > 2$, it is known as the generalized Jordan curve theorem [124, 125]. For technical reasons, one works with the one-point compactifications $\mathbb{S}^n$ of $\mathbb{R}^n$ in algebraic topology; the analogous results for $\mathbb{S}^n$ are known as the Jordan-Brouwer separation theorems [74, 78, 100]. It turns out that boundaries of D-intervals in order complete spaces have the same property.

Light ray segments and timelike curves cannot not have self-intersections; they are necessarily *simple*.[4] However, this is not true of spacelike curves, and one has to make the distinction.

Definition 8.2.2 (Spacelike curves)

Let M be an order complete space. A *spacelike curve* in M is a continuous map $\mathcal{C} : [0,1] \to M$ such that any two distinct points $\mathcal{C}(t)$, $\mathcal{C}(t')$ on it are spacelike to each other. A spacelike curve is called *simple* if it is, additionally, a simple curve, i.e., if it is a homeomorphism onto the co-domain. $\square$

Theorem 8.2.3 *Let M be an order complete space, $U \subset M$ a connected D-set and $I[a,b] \subset U$ a closed D-interval in U, with $a \ll b$.*

1. *$\partial I[a,b]$ is the common boundary of $I(a,b)$ and $U \setminus I[a,b]$.*
2. *Any simple curve that exits the interior of $I[a,b]$ intersects the boundary of $I[a,b]$.*

Proof:

1. Let $x \in \partial I[a,b]$, and let l_x^{out} be a (forward or backward) light ray from x out of $I[a,b]$. Then x is a limit point of $l(x,v) \subset (U \setminus I[a,b])$ (or $l(u,x) \subset (U \setminus I[a,b])$, as the case may be).
2. Let $\mathcal{C}[p,q]$ be a simple curve such that $p \in I(a,b)$ and $q \in U \setminus I[a,b]$. Let $x \in X = \mathcal{C}[p,q] \cap I(a,b)$ and $z \in Z = \mathcal{C}[p,q] \setminus I[a,b]$. Define now

$$Y = \bigcap_{\substack{x \in X \\ z \in Z}} \Theta[x,z] \ .$$

Then Y is nonempty and consists of a single point, $Y = \{y\}$, and $y \in \partial I[a,b]$. $\blacksquare$

---

[4] Since a curve is, by definition, a continuous image of the real interval $[0,1]$, a light ray segment in an order complete space is a curve. However, nothing in our axioms prevents light rays or timelike continua from being homeomorphic with the long line.

Observe that the above proof will fail unless U is order complete (cf. Sect. 7.1). The result is readily extended to the one-point compactification of M, if M is locally compact. We omit the details. Regarding the analogy with the Jordan curve theorem or with the Jordan-Brouwer theorems, one should bear in mind that not every double cone in M is a D-interval.

8.2.2 2-cells in D-intervals

Let M be an order complete space, $I[a, b] \subset M$ a D-interval, $\Theta[a, b]$ a timelike curve connecting a with b, and p a point on $S(a, b)$. There exists a natural homeomorphism between $l[a, p]$ and $\Theta[a, b]$, which may be displayed as follows. Define

$$\vartheta : l[a, p] \to \Theta[a, b] \tag{8.2}$$

by

$$r \in l[a, p], \quad \vartheta(r) = \beta C_r^- \cap \Theta[a, b] . \tag{8.3}$$

By Lemma 8.2.1, the point $\vartheta(r)$ exists and is unique. The map ϑ is clearly bijective and order-preserving, and therefore a homeomorphism. Define now

$$F[a, b; p] = \bigcup_{r \in l[a, p]} l[r, \vartheta(r)] . \tag{8.4}$$

Our aim in this section is to prove that $F[a, b; p]$ is a 2-*cell*, i.e., a homeomorphic image of a 2-simplex. We shall use the following notations:

$$\partial F[a, b; p] = l[a, p] \cup l[p, b] \cup \Theta[a, b]$$
$$F(a, b; p) = F[a, b; p] \setminus \partial F[a, b; p] . \tag{8.5}$$

The notation $\partial F[a, b; p]$ will be justified below.

We begin by observing that if $z \in F[a, b; p]$, then the two intersections

$$\eta = \beta C_z^+ \cap \Theta[a, b] ,$$
$$\zeta = \beta C_z^- \cap \Theta[a, b] \tag{8.6}$$

are uniquely defined points, with $\eta = \zeta$ iff $z \in \Theta[a, b]$. Since $\Theta[a, b]$ is homeomorphic with $[0, 1]$, it is metrizable. We choose a metric on $\Theta[a, b]$, denoted by $|., .|$, so that $|a, b| = 1$. Using this metric, we set up two different ways of identifying a point on $\Theta[a, b]$ (Fig. 8.4):

1. By its distance from the vertex b. η^α will denote the point on $\Theta[a, b]$ at the distance α from b.
2. By its distance from the vertex a. ζ^β will denote the point on $\Theta[a, b]$ at the distance β from a.

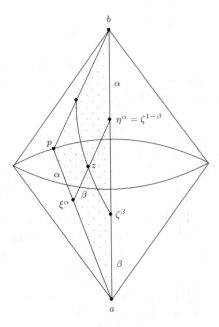

Fig. 8.4. 2-cells in order intervals

Then $0 \leq \alpha, \beta \leq 1$. For any point on $\Theta[a, b]$, $\alpha + \beta = 1$, so that $\eta^\alpha = \zeta^{1-\beta}$. The point $z \in F[a, b; p]$ will be assigned the coordinates

$$z = (\alpha, \beta) , \qquad (8.7)$$

where (see Fig. 8.4) α and β are defined by the intersections

$$\begin{aligned} \eta^\alpha &= \beta C_z^+ \cap \Theta[a, b] , \\ \zeta^\beta &= \beta C_z^- \cap \Theta[a, b] . \end{aligned} \qquad (8.8)$$

Note that in (8.6), z stands for the generic point in $F[a, b; p]$ whereas the left-hand sides of (8.8) are defined for a specific z.

We now set

$$\xi^\alpha = \beta C_{\eta^\alpha}^- \cap l[a, p] . \qquad (8.9)$$

It follows from the definition of $F[a, b; p]$ that

$$z \in l_{\xi^\alpha, \eta^\alpha} .$$

In the following, we shall use the shorthand notation l^α to denote the light ray $l_{\xi^\alpha, \eta^\alpha}$.

The coordinate lines $\alpha = $ const are clearly the light ray segments $l^\alpha[\xi^\alpha, \eta^\alpha]$. On $l^\alpha[\xi^\alpha, \eta^\alpha]$, β varies from 0 to $1 - \alpha$. From (8.8) we see that the coordinate lines $\beta = $ const are the sets

$$\mathcal{C}_\beta = F[a, b; p] \cap \beta C^+_{\zeta^\beta} . \tag{8.10}$$

On $\mathcal{C}_\beta$, α varies from 0 to $1 - \beta$.

Equations (8.7) and (8.8) define an injective map $\Pi' : F[a, b; p] \to \mathbb{R}^2$. With the standard Cartesian coordinates on $\mathbb{R}^2$, we see that Π' maps $F[a, b; p]$ onto the triangle Δ with vertices $(0, 0), (1, 0)$ and $(0, 1)$. That is, the map

$$\Pi : F[a, b; p] \to \Delta , \tag{8.11}$$

which is the same as Π', but with the co-domain restricted to Δ, is bijective. It should, however, be noted that the length scales on $l[a, p]$ and $\Theta[a, b]$ are not identical. Therefore, after the coordinates have been assigned, we should "forget" the scale on $\Theta[a, b]$.

Let $\alpha \in (0, 1)$ be fixed, and $\delta > 0$ such that $\alpha \pm \delta \in (0, 1)$. Then $\cap_\delta l[\xi^{\alpha+\delta}, \xi^{\alpha-\delta}] = \xi^\alpha$ and $\cap_\delta \Theta[\eta^{\alpha+\delta}, \eta^{\alpha-\delta}] = \eta^\alpha$. Furthermore, $\xi^{\alpha+\delta} \ll \eta^{\alpha-\delta}$ for all δ.

Lemma 8.2.4

$$\bigcap_\delta I[\xi^{\alpha+\delta}, \eta^{\alpha-\delta}] = l[\xi^\alpha, \eta^\alpha] .$$

Proof: Let $L = \cap_\delta I[\xi^{\alpha+\delta}, \eta^{\alpha-\delta}]$. As $l[\xi^\alpha, \eta^\alpha] \subset I[\xi^{\alpha+\delta}, \eta^{\xi-\delta}]$ for all δ, it follows that

$$l[\xi^\alpha, \eta^\alpha] \subset \bigcap_\delta I[\xi^{\alpha+\delta}, \eta^{\alpha-\delta}] .$$

We shall assume that the intersection L contains a point $w \notin l[\xi^\alpha, \eta^\alpha]$, and derive a contradiction.

Since M is regular, the closed set $l[\xi^\alpha, \eta^\alpha]$ can be separated from the point w by disjoint open sets $V \supset l[\xi^\alpha, \eta^\alpha]$ and $W \ni w$. The light ray segment $l[a, p]$ and the timelike curve $\Theta[a, b]$ intersect V in an open light ray segment and an open timelike arc respectively, that is the intersections $l(a, \xi^\alpha) \cap V$ and $\Theta(\eta^\alpha, b) \cap V$ are nonempty. Therefore one can find a pair of points $\xi^{\alpha+\delta_0} \in l[a, p]$ and $\eta^{\alpha-\delta_0} \in \Theta(\eta^\alpha, b)$ such that $I[\xi^{\alpha+\delta_0}, \eta^{\alpha-\delta_0}] \subset V$, i.e, $I[\xi^{\alpha+\delta_0}, \eta^{\alpha-\delta_0}]$ is disjoint from W. Therefore $w \notin L$, which is the desired contradiction. ∎

With the help of this lemma, we can prove the following result:

Theorem 8.2.5 *The subset $F[a, b; p] \subset U$ defined in (8.4) has the following properties:*

1. *Let $z \in F[a, b; p]$ and $u, v \in I(a, b)$ such that $u \ll z \ll v$ Then the intersections $F[a, b; p] \cap I(u, z)$ and $F[a, b; p] \cap I(z, v)$ are nonempty.*
2. *Every ray segment $l^\alpha[\xi^\alpha, \eta^\alpha] \subset F[a, b; p]$ separates $F[a, b; p]$ into two disjoint regions that have $l^\alpha[\xi^\alpha, \eta^\alpha]$ as their common boundary.*
3. *No proper subsegment of any $l^\alpha[\xi^\alpha, \eta^\alpha]$ separates $F[a, b; p]$ into two disjoint regions.*

Proof:

1. Let $z \in F(a, b; p)$, $u, v \in I(a, b)$ and $u \ll z \ll v$. Then $z \in l^{\alpha_0}(\xi^{\alpha_0}, \eta^{\alpha_0})$ for some α_0. We shall prove that for any such v, there is a ray segment $l^{\alpha}(\xi^{\alpha}, \eta^{\alpha})$ that intersects $I(z, v)$.
   Let $\eta^{\alpha_1} = \beta C_v^+ \cap \Theta(a, b)$. Then $z \ll \eta^{\alpha_1}$, $\alpha_1 < \alpha_0$ and $\eta^{\alpha_1} \gg \eta^{\alpha_0}$. Consider the D-interval $I[z, \eta^{\alpha_1}]$ and choose α such that $\alpha_1 < \alpha < \alpha_0$. Then $\eta^{\alpha} \in I(z, \eta^{\alpha_1})$, so that every backward ray from η^{α} intersects βC_z^+. This includes the ray segment $l[\xi^{\alpha}, \eta^{\alpha}]$, where $\xi^{\alpha} = \beta C_{\eta^{\alpha}}^- \cap l[a, p]$, as before. Note that $\xi^{\alpha_0} <^{ll} \xi^{\alpha}$, and therefore ξ^{α} is spacelike to z.
   Mutatis mutandis, the proof applies to the statement that there is a ray segment $l(\xi^{\alpha'}, \eta^{\alpha'})$ that intersects $I(u, z)$.
2. For $\xi^{\alpha}, \eta^{\alpha}$ as in the statement of the theorem, the two regions of $F[a, b; p]$ are $F[a, b; p] \cap C_{\xi^{\alpha}}^+$ and $F[a, b; p] \cap C_{\eta^{\alpha}}^-$. The intersection $l[\xi^{\alpha}, \eta^{\alpha}]$ of the cones $C_{\xi^{\alpha}}^+$ and $C_{\eta^{\alpha}}^-$ is the common boundary of these two regions.
3. It is easy to see (for example) that there is an l-polygon, or a segment of the timelike curve $\Theta[a, b]$, or a concatenation of the two, connecting any two points of $F[a, b; p] \setminus l^{\alpha}[\xi^{\alpha}, z^{\alpha}]$ for any α and any $z <^{ll} \eta^{\alpha}$.   ∎

The second part of this theorem establishes that $F[a, b; p]$ is, in common parlance, a continuous surface.

Lemma 8.2.6 *The sets C_{β} defined by* (8.10) *are continuous simple curves.*

Proof: C_{β} is the image of an injective map from $l[\xi^{1-\beta}, \xi^0]$, in coordinates $[0, 1 - \beta]$, into $F[a, b; p]$, and this map is continuous. For suppose that it is discontinuous at $\alpha_0 \in [0, 1 - \beta]$. Then the images of either the sections $[0, \alpha_0]$, $(\alpha_0, 1 - \beta]$ or the sections $[0, \alpha_0)$, $[\alpha_0, 1 - \beta]$ do not meet. In either case, $l[\xi^{\alpha}, \eta^{\alpha}]$ is not the common boundary of the two regions of $F[a, b; p]$ defined by $\alpha < \alpha_0$ and $\alpha > \alpha_0$, which contradicts Theorem 8.2.5. C_{β} is simple because $\beta C_{\zeta^{\beta}}^+$ intersects each light ray segment $l^{\alpha}[\xi^{\alpha}, \eta^{\alpha}]$ at most once.   ∎

With this preparation behind us, we are finally able to reach the goal of this section, which is to prove that the map Π defined earlier is a homeomorphism. The topology on $F[a, b; p]$ is the subspace topology inherited from M; that on Δ is the subspace topology inherited from $\mathbb{R}^2$.

Theorem 8.2.7 *The map $\Pi : F[a, b; p] \to \Delta$ defined by* (8.11) *is a homeomorphism, i.e., $F[a, b; p]$ is a 2-cell.*

Proof: Define

$$R = \{z = (\alpha, \beta) | \alpha_1 < \alpha < \alpha_2, \beta_1 < \beta < \beta_2\},$$

where $\alpha_1, \alpha_2, \beta_1, \beta_2 \in (0, 1), \alpha + \beta < 1$, and call it an "open rectangle" in $F(a, b; p)$. Next, let U_F be the (nonempty) intersection of a D-interval U with F. It is easily seen that given R and U_F such that $R \supset U_F$, one can find an open rectangle $R' \subset F(a, b; p)$ such that $R \supset U_F \supset R'$. This means that the family of open rectangles is a base for the topology of $F(a, b; p)$.

By definition, Π and Π^{-1} map open rectangles in $F(a, b; p)$ bijectively to open rectangles in Δ and vice versa, and boundaries to boundaries. Therefore Π, Π^{-1} restricted to the interiors are homeomorphisms. Therefore they map Cauchy sequences to Cauchy sequences. The result follows.   ∎

8.2.3 Cylindrical Coordinates on D-intervals

In this section we shall establish several decisive results, using a system of cylindrical coordinates on the D-interval $I[a, b]$ for a given timelike curve $\Theta[a, b]$. This coordinate system is defined below.

We begin by mapping $\Theta[a, b]$ on the real interval $[-1, 1]$, so that

$$a \mapsto -1 \,,$$
$$b \mapsto +1 \,. \tag{8.12}$$

Let now $w \in I[a, b]$, $u = C_w^- \cap \Theta(a, b)$ and $v = C_w^+ \cap \Theta(a, b)$. We shall use the same letters u and v to denote the real coordinates of the points u and v on $\Theta[a, b]$. Then

$$-1 \le u \le v \le 1 \,.$$

If $w \in \Theta[a, b]$, then $u = v$. If $w = b$, then $u = v = 1$, and if $w = a$, then $u = v = -1$. We define

$$r = \frac{v - u}{2} \,,$$
$$h = \frac{v + u}{2} \,. \tag{8.13}$$

Then

$$-1 \le h \le 1 \,,$$
$$0 \le r \le 1 - |h| \,. \tag{8.14}$$

The variables h and r will be called the *level* and the *radius*, respectively. Note that $r = 0$ implies that $w \in \Theta[a, b]$, and that the upper bound for r is a consequence of $u \le v$.

For fixed h, the sets of constant nonzero r are the hyperspheres $S(u, v)$. To complete the parametrization of $I[a, b]$, we have to parametrize $S(u, v)$. For this we need to know more about $S(a, b)$ than we presently do, for example that $S(a, b)$ is a topological manifold. However, for our immediate aims, which are to establish the homogeneity Property 8.2.9 and the homotopy Property 8.2.11, it suffices to use the second homogeneity Property 5.4.5: If $a, b \in U$ and $a', b' \in U'$, where U, U' are D-intervals, then $S(a, b)$ and $S(a', b')$ are homeomorphic with each other, and therefore to a *fiducial hypersphere* in M which we shall denote by S.

It follows from the definitions of u, v and $F[a,b;p]$ that the intersection $S(u,v) \cap F[a,b;p]$ is a unique point. It is, in fact, the point w that was used to define u and v earlier, i.e.,

$$S(u,v) \cap F[a,b;p] = w .$$

Let $\phi(p)$ be a set of real coordinates that specify the point $p \in S(a,b)$ uniquely. Then, in view of the second homogeneity property, we may demand that for every u and v points on the hyperspheres $S(u,v)$ be assigned coordinates that satisfy the following condition:

$$\phi(w) = \phi(p) \ \forall \ w \in F[a,b;p]. \tag{8.15}$$

In other words, *we require the $\phi(w)$ to be constant on $F[a,b;p]$.* We shall generally write (8.15) in the abbreviated form

$$\phi(w) = \phi(p) = \text{const} \ \forall \ w \in F[a,b;p] . \tag{8.16}$$

Since M is a Tychonoff space (Theorem 4.6.3), we may assume that the functions $\phi(p)$ are continuous in p in any coordinate chart. The (affordable) price one pays is that more than one chart will be needed to cover S.

Thus a point $w \in I[a,b]$ may be specified as

$$w = \{h; r, \phi(w)\} . \tag{8.17}$$

For $r > 0$, the specification is unique. For $r = 0$, the coordinate $\phi(p)$ does not play a role, i.e., all values are equivalent. The parameters h, r and ϕ are continuous functions of w.

Let T be the triangle in the (r,h) plane with vertices $(0,-1)$, $(0,1)$ and $(1,0)$, and S the fiducial hypersphere. Then the parameter space for the parametrization (8.17) of $I[a,b]$ may be written as

$$K = (T \times S)/R, \tag{8.18}$$

where R is an equivalence relation on $T \times S$ that identifies all points $\{h; 0, \phi\}$ with h fixed, but does nothing for $r \neq 0$. Points in $T \times S$ will be denoted by $\{h; r, \phi\}$ as in (8.17), and those on K by $[h; r, \phi]$. Then the map $I[a,b] \to K$ defined by

$$w \mapsto [h; r, \phi(w)] \tag{8.19}$$

will be bijective.

Theorem 8.2.8 *The map $f : I[a,b] \to K$ defined by (8.19) is a homeomorphism.*

Proof: f is continuous because the functions h, r and ϕ are continuous. The proof that f^{-1} restricted to $\{h; r, \phi\}|r > 0\}$ is continuous is similar to the proof of Theorem 8.2.7. The extension to $I[a,b]$ and K is straightforward, if slightly tedious. We omit the details. ∎

8.2.4 Homogeneity and Homotopy of D-intervals

We are now in a position to establish the following key results on D-intervals:

Theorem 8.2.9 (Third homogeneity property: $I[a,b]$)

Let $I[a,b]$ and $I[a',b']$ be D-intervals with nonempty interiors. Then $I[a,b]$ and $I[a',b']$ are homeomorphic to each other.

Proof: Follows from the fact that both $I[a,b]$ and $I[a',b']$ are homeomorphic with K. ∎

Theorem 8.2.10 *Let $I[a,b]$ be a closed D-interval with nonempty interior, and let o be any point in $I(a,b)$. Then $I[a,b]$ is contractible to $\{o\}$.*

Proof: Recall that the concatenation of two timelike curves $\Theta[a,o]$ and $\Theta[o,b]$ is again a (continuous) timelike curve. Map $\Theta[a,o]$ onto $[-1,0]$ and $\Theta[o,b]$ onto $[0,1]$. Then the point $p \in I[a,b]$ can be parametrized as $p = [r, h, \phi]$, with $o = [0,0,\phi]$. The map

$$p = p(1) \mapsto p(t) = [th; tr, \phi] , \tag{8.20}$$

defined for every $t \in [0,1]$, displays the required contractibility. ∎

Note that ϕ remains constant during the contraction. Each point p on $\partial I[a,b]$ is joined to the point o by a continuous curve. The paths of distinct points on $\partial I[a,b]$ meet only at o, and the collection of all these paths fills $I[a,b]$.

It follows easily from Theorem 8.2.10 that:

Theorem 8.2.11 (D-intervals are homotopically trivial)

Let $I[a,b]$ be a closed D-interval. Then $I[a,b]$ is homotopically trivial.

Proof: The result is trivially true if $a = b$. If $a \neq b$ then, since $\exists$ a D-set $U \supset I[a,b]$, we can find points $a_0, b_0 \in U \setminus I[a,b]$ such that $a_0 \ll a \ll b \ll b_0$. Then, by Theorem 8.2.10, $I[a_0, b_0]$ is contractible to any point $o \in I(a_0, b_0)$. The result follows from the fact that $I[a,b] \subset I(a_0, b_0)$. ∎

8.2.4.1 Spacelike Hypersurfaces

Under the contraction (8.20), the path traced by any point $p \in S(a,b)$ is a spacelike curve. If $I[a,b]$ is parametrized as in (8.12) and (8.13), the paths $\{p(t)|p \in S(a,b), \ 0 \leq t \leq 1\}$ lie on the bounded hypersurface $\bar{H} = \{[0; r, \phi]|r \in [0,1], \ \phi \in S(a,b)\}$. We shall denote by H the subset of $\bar{H}$ defined by $H = \{[0; r, \phi]|r \in [0,1), \ \phi \in S(a,b)\}$. Both $\bar{H}$ and H will be called, indifferently, the level surface $h = 0$.

On H, the condition $r = \text{const} = r_0 \in (0,1)$ defines a hypersphere which we shall denote by S_{r_0}. S_{r_0} separates H into an outer and an inner part that have S_{r_0} as their common boundary. Although the result is fairly obvious by now, it is of independent interest,[5] and therefore we shall give a detailed statement and proof.

Theorem 8.2.12 *Let M be an order complete space and $I[a,b] \subset M$, $a \ll b$ a closed D-interval. Let $w \in I[a,b]$ be parametrized as in (8.17), and let $H \subset I(a,b)$ be the level surface $h = 0$. Then:* 1) *For $0 < r_0 < 1$, the hypersphere S_{r_0} separates H into two disjoint parts*

$$H_{\text{in}} = \{[0; \rho, \phi(w)] | \rho < r_0\}$$

and

$$H_{\text{ex}} = \{[0; \rho, \phi(w)] | \rho > r_0\}$$

that have S_{r_0} as their common boundary. 2) *Furthermore, any continuous curve in H that joins $p \in H_{\text{in}}$ with $q \in H_{\text{ex}}$ intersects S_{r_0}.*

Proof: Let $z \in S_{r_0}$, $u \in \beta C_z^- \cap \Theta[a,b]$, $v \in \beta C_z^+ \cap \Theta[a,b]$. Then

$$H_{\text{in}} = H \cap I(u,v),$$

$$H_{\text{ex}} = H \cap (I[a,b] \setminus I[u,v])$$

and

$$S_{r_0} = H \cap \partial I[u,v] \ .$$

This establishes assertion 1).

Assertion 2) is clearly true for the radial lines $\phi = \text{const}$. So let $\mathcal{C} = \mathcal{C}[p,q]$ be a closed segment of a continuous spacelike curve that joins a point $p \in H_{\text{in}}$ with a point $q \in H_{\text{ex}}$. By definition, $\mathcal{C}$ is homeomorphic to a closed interval $[\alpha, \beta] \subset \mathbb{R}$. Let $x \in \mathcal{C}_{\text{in}} = \mathcal{C} \cap H_{\text{in}}$ and $z \in \mathcal{C}_{\text{ex}} = \mathcal{C} \cap H_{\text{ex}}$. The intersection

$$\bigcap_{\substack{x \in \mathcal{C}_{\text{in}} \\ z \in \mathcal{C}_{\text{ex}}}} \mathcal{C}[x,z]$$

consists of a single point y, and $y \in S_{r_0}$.   ∎

8.3 Locally Compact Spaces

In Sect. 6.7 we remarked that many of the results of Chap. 6 could be valid without the assumption of local precompactness. If that were the case, the same would be true of the results of the preceding section. However, the results that follow, leading up to and including the results on the differentiable structure, are of a different kind. As will be seen below, their proofs depend

---

[5] The interest here is the following. It may be possible to establish analytical or topological properties of $\mathbb{R}^n$ or $\mathbb{S}^n$ by imposing an auxiliary partial order structure on these spaces.

critically upon the local compactness condition. The methods used cannot be extended readily to "infinite-dimensional" situations, and therefore it is difficult to see which of the results, if any, may continue to hold. In order to stress this point, the hypothesis of local compactness will be mentioned explicitly in the following.

We shall establish that open D-intervals in locally compact order complete spaces have a differentiable structure by embedding them as open double cones in some finite-dimensional Minkowski space. Our point of departure is the following result:

Theorem 8.3.1 *A locally compact order complete space has a finite covering dimension.*

Proof: In a locally compact space M, a closed D-interval $I = I[a, b]$ is compact. Therefore every open cover of it has a finite subcover. Let $\mathcal{O}$ be an open cover of I and $j(\mathcal{O})$ the size of a finite subcover of it. Let $k + 1 = \min j(\mathcal{O})$, the minimum being taken over all finite subcovers of all open covers of I. Since I is connected, $k \geq 3$.[6] Since every closed D-interval is homeomorphic with every other closed D-interval, k is independent of the D-interval.

The number k is the *covering* or *topological dimension of M.*   ∎

Next, we recall some results from the theory of embedding and metrization of topological spaces:

1. A compact Hausdorff space is metrizable [77].
2. A metrizable space of topological dimension k can be embedded in $\mathbb{R}^n$, where $n = 2k + 1$ [77, 52]. This value of n is the best possible. However:
3. The topological dimension of a compact m-dimensional topological *manifold* is exactly m [77, 52].[7]

The only useful conclusion that we can draw from the above results is that a D-interval in a locally compact order complete space can be embedded in $\mathbb{R}^n$ for some finite n. For n to equal the topological dimension of the D-interval, the result quoted requires the D-interval to be a topological *manifold*, which begs the question. In a manifold, every point has a small neighbourhood that is homeomorphic with a small neighbourhood of every other point, and open D-intervals have this homogeneity property (Property 8.2.9). However, in a finite-dimensional manifold, every point has an open neighbourhood that is homeomorphic with an open set in some $\mathbb{R}^N$, and this is the crucial property that we are still missing.

---

[6]   Recall that we have excluded the one- and two-dimensional cases.

[7]   Recall that a uniform space is metrizable iff it is Hausdorff and second countable (Theorem A.5.4). However, for the moment we are interested in embeddings in $\mathbb{R}^n$ with finite n; for this second countability is not enough. In some sense, the second countability condition is more restrictive globally than locally; Hilbert space, for example, is second countable, but the long line is not. See the articles by Kneser [59, 61].

Let M be a locally compact order complete space of topological dimension k. We take an embedding of M into $\mathbb{R}^{2k+1}$ and the assignment (8.19) of cylindrical coordinates to points $w \in I[a, b] \subset U \subset M$ as described in Sect. 8.2.3. Then:

Lemma 8.3.2 *Let i be an embedding of $I[a, b]$ in $\mathbb{R}^{2k+1}$. Then i maps the paths of points $p \in \partial I[a, b]$ under the homotopy (8.20) onto straight lines.*

Proof: Since ϕ remains constant during a contraction, the path $\{p(t)|t \in [0, 1]\}$ traced by p during the homotopy (8.20) lies on the 2-cell $F[a, b; p]$. Formula (8.20) shows explicitly that i maps the cell $F[a, b; p]$ into the plane in $\mathbb{R}^{2k+1}$ determined by $h(p)$ and $r(p)$, and the path of p onto a straight line in this plane. ∎

8.3.1 Reconstruction of a Local Minkowski Structure

Let o be the point $[0; 0, \phi] \in I[a, b]$ and $p_\gamma \in S(a, b)$. Set $i(o) = O$ and $i(p_\gamma) = P_\gamma$. The points O and P_γ determine a vector, which we shall denote by $\mathbf{v}_\gamma$, in $\mathbb{R}^{2k+1}$. Let

$$V = \{\mathbf{v}_\gamma | p_\gamma \in S(a, b)\} \, ,$$

and set

$$\mathbb{H} = \{a_1 v_1 + a_2 v_2 + \cdots + a_k v_k | a_j \in \mathbb{R}, \, v_j \in V, \, k \in \mathbb{N}\} \, .$$

Then $\mathbb{H}$ is the linear span of the vectors of V and is a subspace of $\mathbb{R}^{2k+1}$, of dimension $\leq 2k$. We shall call it the *hyperplane* $\mathbb{H}$. Set $\dim \mathbb{H} = N - 1$. It follows immediately that:

$$\dim i(I[a, b]) = N \, . \tag{8.21}$$

Then the homogeneity Property 8.2.9 of the interior of $I[a, b]$ implies that $N = k$, the topological dimension of M. Thus $I(a, b)$ is embedded as an open set in $\mathbb{R}^N$.

We now define the indefinite form $x_0^2 - x_1^2 - \cdots - x_{N-1}^2$ on $\mathbb{R}^N$ to make it into N-dimensional Minkowski space, which we shall denote by $\mathbb{M}$. Let A and B be the points $(-1, 0, \ldots, 0)$ and $(1, 0, \ldots, 0)$ respectively in $\mathbb{M}$, and $I_\mathbb{M}[A, B]$ the double cone $C_A^+ \cap C_B^-$. The intersection $\partial C_A^+ \cap \partial C_B^-$ is the sphere $\mathbb{S}^{N-2}(1; O)$ of unit radius with origin O which lies on the hyperplane $x_0 = 0$. We introduce cylindrical coordinates on $\mathbb{R}^N$ with the x_0-axis as the axis of the cylinder and O as the origin. We shall denote a point in $\mathbb{M}$ by $[h; r, \phi]_\mathbb{M}$, where h and r are the standard level and radius variables, and ϕ specifies a point on an Euclidean hypersphere $\mathbb{S}^{N-2}$. The assignment of the coordinate ϕ satisfies the condition satisfied by ϕ in (8.15). The map

$$[h; r, \phi] \mapsto [h; r, \phi]_\mathbb{M} \tag{8.22}$$

is a homeomorphism of $I[a, b] \subset M$ onto its range in $I_\mathbb{M}[A, B]$.

8.4 The Differentiable Structure

Theorem 8.4.1 *The interior of a D-interval in an order complete space M is C^∞-diffeomorphic with $\mathbb{R}^N$, where N is the dimension of M.*

Proof: This is just a restatement of the known fact that the interior of a double cone in $\mathbb{M}^N$ carries a C^∞-differential structure.   ∎

Example 8.4.2 Consider two copies M and M' of the plane $\mathbb{R}^2$, with its usual differentiable structure. Let M be furnished with the usual Minkowski structure.[8] Define the map $f : M \to M'$ as follows:

$$f(x,y) = \begin{cases} (x,y), & y \leq 0 ; \\ (x + \alpha y, y), & y > 0 , \end{cases}$$

where $\alpha > 0$. This map is a homeomorphism. If we define light rays in M' to be the images under f of light rays in M, the space M' becomes an ordered space, and the spaces M and M' are order-equivalent (as defined in Sect. 4.7). However, light rays in M' are not differentiable where they intersect the X-axis.

This example reveals that the notion of order equivalence is not a diffeomorphism invariant.

We have to distinguish between the differentiability of a manifold, and of the order structure upon it. We begin with a definition that makes the latter concept precise.

Definition 8.4.3 Let M be an ordered space that is also a differentiable manifold. We shall say that the order structure on M is differentiable if the differentiable structure on M induces a differentiable structure on (i) light rays, (ii) mantles of light cones (excluding the vertices), and (iii) spacelike hyperspheres in D-sets.   □

We saw in Theorem 8.4.1 that every open D-interval in a locally compact order complete space M carries a differentiable structure. However, this does not imply that the differentiable structures on two open D-intervals in M with non-empty intersection are compatible.[9] It cannot be expected that in a setting as general as the present one (of order complete spaces) a local differentiable structure would imply a global one, or that the order structure itself would be differentiable everywhere.

---

[8] In two-dimensional Minkowski space, light rays are the straight lines $x - y = c$ and $x + y = c$, where $c \in \mathbb{R}$.

[9] Here compatibility has the usual meaning; the transition maps for two overlapping open D-intervals are related by a diffeomorphism on their intersection.

8.4.1 Remarks on the Global Differentiable Structure

In Example 8.4.2, an everywhere-differentiable order structure is mapped, by a order preserving homeomorphism, into one in which light rays are not differentiable at their intersections with the X-axis. Further insight into the phenomenon may be obtained by looking at the standard maps ρ and σ which were defined in Sect. 5.1.1, and which formed the basis of the subsequent investigations. In Example 8.4.2, the induced maps $\rho' = \rho \circ f$ and $\sigma' = \sigma \circ f$ on $\mathbb{M}'$, while being continuous, are not differentiable if one of the ray segments crosses the X-axis.

In any order complete space, a light ray segment is homeomorphic with an interval on the real line. Therefore the notion of differentiability itself is always meaningful for the standard maps. A failure of differentiability of the order structure should therefore manifest itself as a failure of differentiability of the standard maps, and this would be an *intrinsic* property of the space under consideration. Conversely, (it seems reasonable to suggest that) differentiability of the standard maps should allow one to infer the differentiability of the order structure, and perhaps that of the space itself.

Despite the obstructions for defining a global differentiable structure, there are indications that one is not very far from such a structure. These indications are as follows:

1. In Sect. 4.6 we established that an ordered space is a Tychonoff space (Theorem 4.6.2, Corollary 4.6.3), which means that one can define continuous functions on it which separate points. If the space is locally compact, we may construct the C^*-algebra generated by the continuous functions with compact support. The space itself is characterized by this C^*-algebra.[10] By this construction we obtain all continuous function vanishing at ∂M and at infinity. Then one can characterize the order structure by monotone functions, which can be constructed from the continuous functions. Monotone functions on the line are differentiable almost everywhere (see, for instance, [89]). Since the space is finite dimensional, one can define the tangent space almost everywhere.

2. Let M be a locally compact order complete space and $I[x, y] \subset M$ a D-interval in it. Let $\Theta(t)$, $t \in [0, 1]$ be a parametrized timelike curve joining x and y, with $\Theta(0) = x$ and $\Theta(1) = y$. Let $a \in S(x, y)$, and define $a(t) = \beta C_{\Theta(t)}^- \cap l_{x,a}^+$. Embedding M in an n-dimensional Euclidean space, we can define $\frac{1}{t}(a(t) - a(0))$. Taking the limit $t \to 0$ and varying a over $S(x, y)$, we may define a family of "one-sided tangent vectors" at x, along the directions of light rays from x. Therefore it seems reasonable to assume the existence of tangent spaces and cones in it.

---

[10] This is the famous Gelfand isomorphism. For an elementary account of the basic Gelfand theory, see [99]. More advanced accounts and further material may be found in [24] and [81].

3. In physics, one can make only a finite number of experiments. The data they would provide would be insufficient to distinguish between a space of continuous but nowhere-differentiable functions and a $C^{(\infty)}$ space. Therefore, from the point of view of physics, no real harm can be done by assuming the existence of tangent spaces at every point.

8.4.2 Monotone Functions

We begin with the construction of real-valued functions that are monotonic with respect to the order $<$.[11]

Definition 8.4.4 Let M be an order complete space and $f : M \to \mathbb{R}$ a function on M. For given f, we define a monotone nondecreasing function as follows. Let $a \in M$, and define

$$f^m(a) = \sup\{b \in C_a^- \mid f(b)\} \, .$$

For $a < c$ this implies (by definition of the order $<$) $C_a^- \subset C_c^-$, it follows that the function f^m is monotone nondecreasing.   □

One can define monotone nonincreasing functions in the same manner.

Proposition 8.4.5 *Let M be order complete and locally compact and let $f(a)$ be a continuous function with compact support A. Then the function $f^m(a)$ defined above is continuous.*

Proof: Let $f^m(x)$ a monotone increasing function on M and $I[a, b]$ an order interval. Then $f^m(a) \le f^m(y) \le f^m(b)$ for $y \in I[a, b]$. Therefore to prove that $f^m(x)$ is continuous at x it would suffice to show that for every $\epsilon > 0$ there exists $a \in \tau C_x^-$ such that $f^m(a) \ge f^m(x) - \epsilon$ and $b \in \tau C_x^+$ such that $f^m(b) \le f^m(x) + \epsilon$.

Let D be a D-set and $a, x, b \in D$ such that $a \ll x \ll b$. There exists a timelike curve $\Theta(t)$, $-1 \le t \le 1$ connecting a with b such that $\Theta(0) = x$. For every $y \in \partial C_x^- \cap A$, let U_y be a neighbourhood of y such that $|f(y) - f(z)| < \frac{\epsilon}{2}$ for $z \in U_y$. Since A, the support of $f(z)$, is compact, $\partial C_x^- \cap A$ is also compact and we can find a finite subcover $\{U_i\}$ of $\partial C_x^- \cap A$. Denote by U the union of the sets U_i. Let $t^+ > 0$ and $t^- < 0$ such that $\partial C_{\Theta(t^\pm)}^- \cap A \subset U$. The difference $\{A \backslash U\} \cap C_x^-$ belongs to all three sets $C_{\Theta(t_j)}^- \cap A$, where $t_j = t^+$, 0 or t^-. This implies that

$$\sup\{f(z); z \in C_{\Theta(t^-)}^- \cap A\} + \epsilon \le \sup\{f(z); z \in C_x^- \cap A\}$$

$$\le \sup\{f(z); z \in C_{\Theta(t^+)}^- \cap A\} - \epsilon \, .$$

From this it follows that $f^m(\Theta(t^-)) + \epsilon \le f^m(x) \le f^m(\Theta(t^+)) - \epsilon$. The two inequalities imply that for $z \in I(\Theta(t^-), \Theta(t^+))$

---

[11] The order $<$ on M was defined in Chap. 3, Def. 3.2.6, p. 24.

$$f^m(x) - \epsilon \leq f^m(z) \leq f^m(x) + \epsilon .$$

This proves the proposition.

■

Remark 8.4.6 If we form the C^*-algebra of bounded continuous functions generated by the monotone functions defined above, we obtain a compactification of M. This compactification differs from all the usual ones, because the functions behave differently in spacelike and timelike directions.

8.4.3 Isotropy

We shall not discuss the construction of directions of the light rays at a point $x \in M$ any further. From now on we shall assume that at every point $x \in M$ there exists a tangent space T_x and that the directions of the light rays at x generate the mantle of a cone. Because of the convexity axiom this cone will be strictly convex,[12] meaning that its faces are one-dimensional. If we construct the cone mantle as described at the beginning, i.e, with help of a timelike curve, then the detailed structure of the cone will depend on the direction of the timelike curve at x and the time-scale on this curve. This means that the cone at x is unique up to a linear transformation.

However, this does not imply that the mantle of the cone is smooth. Some further conditions are required for this mantle to be differentiable. One such condition is the physically motivated one of local isotropy.

Definition 8.4.7 A finite dimensional order complete space M on which a tangent space T_x exists at every point $x \in M$ will be said to fulfil the condition of *isotropy* if on each of these tangent spaces there exists a linear representation of the rotation group mapping the trace of $C_x \subset T_x$ onto itself. By *trace* of C_x is meant the cone in T_x discussed above. □

Introducing suitable coordinates in T_x we may give the trace of the cone $C_x \subset T_x$ the form $\xi_0^2 - \sum \xi_i^2 \geq 0$. This implies that every tangent space is covariant under the Lorentz group extended by dilatations. Hence M carries a G-structure as defined in Sect. B.2. Therefore M is a differentiable manifold with a pseudo-Riemannian metric.[13]

---

[12] The theory of topological vector spaces is developed in textbooks like [91] and monographs like [92].

[13] Since the tangent space exists at every point, M must be once-differentiable, which implies that it is a $C^{(\infty)}$ manifold (see, for example, [50]).

8.5 Covering Spaces as Ordered Spaces

When light rays are locally homeomorphic with $\mathbb{R}$, one may study the extension of the order structure if M has a nontrivial fundamental group[14] π_1 and one passes to a covering space.

An ordered space with nontrivial π_1 can be constructed by excising closed sets from an ordered space with trivial π_1, and this is the picture which we shall have in mind. Not all spaces with trivial π_1 may be able to carry an order structure. For example, consider the standard order structure on $\mathbb{R}^2$. This structure does not survive the one-point compactification. In general, we would not expect a compact space to carry an order structure as defined in this work.

On the other hand, a covering space may be given an order structure that does not project to an order structure on the base space. Consider the punctured plane, and define "rays" to be the radii and the circles around the origin. This structure is locally an order structure, but fails globally because circles are closed curves. The universal covering of the punctured plane opens these circles into lines without ends, and therefore the lift of the structure to the universal covering is an order structure.

Since a covering space is a bundle with a discrete group (and fibre), the notion of fibre bundles[15] provides a convenient approach to the problem.

8.5.1 The Coordinate Transformations

Let X be a topological space and B a covering of it. Let $p : B \to X$ be the projection. Then B is a bundle with base X, projection p and group G, where G is a subgroup of $\pi_1(X)$ (with the discrete topology). It follows that there exists a system of coordinate transformations on X with values in G (see Sect. B.1), i.e., an open cover $\mathcal{O} = \{V_j\}_{j \in J}$ of X by D-sets and continuous maps

$$g_{ji} : V_i \cap V_j \to G \qquad (8.23)$$

which satisfy the conditions (B.2) and (B.3) of Appendix B.

Let W be a connected subset of $V_i \cap V_j$. Then, since G is discrete, $g_{ji}(x)$ is constant on W:

$$g_{ji}(x_1) = g_{ji}(x_2) \quad \forall x_1, x_2 \in W . \qquad (8.24)$$

Let now "$<$" be an order relation on X, i.e., let $(X, <)$ be an ordered space, and consider the cover $\mathcal{O}$ of X. Each $V_i, i \in \mathcal{O}$, becomes an ordered space $(V_i, <)$ by restriction from $(X, <)$, and if $(V_i, <)$, $(V_j, <)$ are two such ordered open subsets of $(X, <)$, then their order structures agree, by definition, on the intersection $V_i \cap V_j$.

---

[14] Arbitrary continuous curves in M can be constructed by the concatenation of timelike and spacelike curves, and l-polygons.

[15] The definition of fibre bundles is given in Appendix B.

If $B = X$, then $G = \{e\}$ and the maps (8.23) become

$$g_{ij}(x) = e \quad \forall\, x \in V_i \cap V_j, \ \forall\, i,j \in J\,.$$

In the general case, $p^{-1}(V_i)$ consists of $|G|$ identical copies of V_i. We introduce a coordinate system on X with values in G as follows: Define a new indexing set $A = J \times G$, and define $V_{i,g}$ by

$$V_{i,g} = V_i \quad \forall\, g \in G\,.$$

For $V_{i,g} \cap V_{j,g} \neq \emptyset$, define the coordinate transformations

$$g_{j,g';i,g} : V_{i,g} \cap V_{j,g'} \to G \tag{8.25}$$

as follows:

$$g_{j,g';i,g}(x) = g^{-1}g' \tag{8.26}$$

for all $x \in V_{i,g} \cap V_{j,g'}$. It is clear that the transformations so defined are continuous, and that they satisfy the equations (B.2) and (B.3). To sum up:

Theorem 8.5.1 *Let $(X, <)$ be an ordered space, and $\mathcal{O} = \{V_j\}_{j \in J}$ an open cover of X. If $(V_i, <)$ is the restriction of $(X, <)$ to V_i, and G is a discrete topological group, then the maps (8.25) and (8.26), which are constant on connected components, form a system of order-preserving coordinate transformations on $(X, <)$ with values in G.* ■

We shall now rename the indexing set $J \times G$ and the double index (j, g) as A and α respectively, for a more convenient notation.

8.5.2 The Total Space

As above, G will be a subgroup of π_1 with the discrete topology. G will also be the fibre, but it will be useful to adopt a distinctive notation. We shall denote the fibre by Y, with $y \in Y$. The group G will act upon the fibre by left-translations.[16]

Endow the indexing set A with the discrete topology, and construct the topological space $X \times Y \times A$. Let $T \subset X \times Y \times A$ be the subset of the triples (x, y, α) such that $x \in V_\alpha$. Then T is a topological space (with the subspace topology), and is the union of disjoint open subsets $V_\alpha \times y \times \alpha$ homeomorphic with V_i for some $i \in J$. Clearly these homeomorphisms are order isomorphisms.[17]

Now define in T an equivalence relation

$$(x, y, \alpha) \sim (x', y', \alpha') \tag{8.27}$$

---

[16] The material of this section is taken from Steenrod's book, [102], Sect. 3.

[17] An *order isomorphism* is another, slightly shorter term for an order-preserving homeomorphism.

if

$$x = x', \qquad g_{\alpha\alpha'}(x) \cdot y = y' . \tag{8.28}$$

The fact that (8.27), (8.28) define an equivalence relation follows from (B.2).

Define B to be the set of equivalence classes of this relation in T, and let

$$q : T \to B \tag{8.29}$$

be the map which sends (x, y, α) in T to its equivalence class $\{(x, y, \alpha)\}$ in B. Finally, let $U \subset B$ be open iff $q^{-1}(U)$ is open in T. Then B becomes a topological space, and q a continuous map.

Define $p : B \to X$ by

$$p(\{(x, y, \alpha)\}) = x . \tag{8.30}$$

(B.3) ensures that p is well-defined. It is shown in [102] that p is continuous.

The coordinate function ϕ_α is defined by

$$\phi_\alpha(x, y) = q(x, y, \alpha), \qquad x \in V_\alpha, y \in Y .$$

The continuity of q implies the continuity of ϕ_α. Equation (8.30) gives $pq(x, y, \alpha) = x$, and therefore $p\phi_\alpha(x, y) = x$. Thus ϕ_α maps $V_\alpha \times Y$ into $p^{-1}(V_\alpha)$. It is shown in [102] that ϕ_α is a (fibre-preserving) homeomorphism.

Finally, one can prove that the $\{g_{\alpha\alpha'}\}$ are the coordinate transformations of the bundle constructed, by showing that

$$\phi_{\alpha',x}^{-1}\phi_{\alpha,x}(y) = g_{\alpha'\alpha}(x) \cdot y \quad \forall\, y \in Y .$$

For details, see [102].

8.5.3 The Order Structure

We may verify that the covering space construction preserves the order structure. The fact that it is preserved locally results from:

Proposition 8.5.2 *Let $U \subset X$ be open in X such that $\pi_1(U) = e$ (U has the subspace topology). Then $p^{-1}(U)$ consists of disjoint subsets of X which are homeomorphic and order-isomorphic with U.*

Proof: Since $\pi_1(U) = e$, the portion of the bundle over U, considered as a bundle in its own right, is the product $U \times Y$. Each section of it is homeomorphic to U. This homeomorphism can be exhibited as

$$x \leftrightarrow (x, g), \quad x \in U, \ g \text{ a fixed element of } G ,$$

which displays that it is also an order-isomorphism. ∎

This establishes that B is locally an ordered space. If light rays are globally and not just locally homeomorphic with $\mathbb{R}$ (which will always be true if X is second countable), then the global part of the order axiom holds in B if it holds in X. This is proven below in Proposition 8.5.3. Since $\mathbb{R}$ is paracompact and contractible, and its fundamental group is trivial, the same would be true of the light rays.

Proposition 8.5.3 *Assume that light rays are globally homeomorphic with* $\mathbb{R}$, *and let l be a light ray in M. Then its lift $\tilde{l}$ to B consists of $|G|$ disjoint light rays.*

Proof: Consider $p^{-1}(l)$, the portion of the bundle B over l. The topological space l is paracompact and contractible, and therefore every bundle based on it equivalent to the product. Therefore $p^{-1}(l)$ is homeomorphic with $l \times |G|$, i.e., it consists of $|G|$ disjoint copies of l. ∎

It follows from Proposition 8.5.3 (covering spaces lift light rays to disjoint light rays) that the covering space construction does not admit of the possibility of "bringing back" a forward light ray to a neighbourhood of a backward point on it. Such situations may arise in constructions such as compactification. They would then lead to a violation of the cone axiom, $C_x^+ \cap C_x^- = \{x\}$. This does not happen in the lifting of the order structure to a covering space. This concludes our demonstration that the order structure on X can be extended to its covering spaces.

Note, finally, that every path can be approximated by light-ray polygons. Therefore the fundamental group π_1 can be defined in terms of light ray polygons. This implies, inter alia, that the covering space is l-connected.

9

The Cushion Problem

9.1 Statement of the Problem

The *cushion problem* is the following. Let M be an order complete space,[1] $U \subset M$ a D-set, $a, b \in U$, $a \ll b$ and $p, q \in \beta I[a, b]$ such that a, b, p, q are not coplanar, for example as in Fig. 9.1. Let $\xi \in l[a, p]$ and $\eta \in l[a, q]$. Then there exist points $\xi' \in l[q, b]$ and $\eta' \in l[p, b]$ such that $\lambda(\xi, \xi')$ and $\lambda(\eta, \eta')$. The question is:

Problem 9.1.1 (The cushion problem) *Do the light rays $l_{\xi,\xi'}$ and $l_{\eta,\eta'}$ intersect each other in U?*

The problem is so named because, if the answer to it is in the negative, the "ribbon" of light rays from $l[a, p]$ to $l[q, b]$ forms a *cushion* with the ribbon of light rays from $l[a, q]$ to $l[p, b]$. The answer to the question is not known in the general case; it is, however, known is that there are no cushions in Minkowski or de Sitter spaces.

Section 9.2 gives a proof of the latter assertion. It is followed by one that discusses the construction of timelike curves in cushion-free spaces. As will be seen, this construction is much simpler than the one given in Sect. 8.1.

9.2 Minkowski and de Sitter Spaces

We begin by showing that, in Minkowski 3-space $\mathbb{M}^3$ defined by the inner product $(x, y) = x_0 y_0 - x_1 y_1 - x_2 y_2$, the four vertices of a light-ray quadrilateral

---

[1] In [11], the cushion problem was encountered while trying to construct timelike curves in which light rays were locally homeomorphic with $\mathbb{R}$, but M itself was not order complete. The problem will also arise in spaces that are locally $\mathbb{F}^n$, where $\mathbb{F}$ is the minimal real algebraic extension of $\mathbb{Q}$ that is closed under taking square roots (see Example 5.2.2 and the footnote to it, p. 58).

H.-J. Borchers and R.N. Sen: *Mathematical Implications of Einstein–Weyl Causality*,
Lect. Notes Phys. **709**, 129–135 (2006)
DOI 10.1007/3-540-37681-X_9
© Springer-Verlag Berlin Heidelberg 2006

lie on a one-sheeted hyperboloid (assuming the four points are not coplanar). To do so, we shall make use of the notion of vector product in $\mathbb{M}^3$. If a, b are two vectors in $\mathbb{M}^3$, their vector product $[a, b]$ is defined to be

$$[a, b] = \{a_1 b_2 - a_2 b_1, -(a_2 b_0 - a_0 b_2), -(a_0 b_1 - a_1 b_0)\} . \tag{9.1}$$

The changes of sign (with respect to the vector product in $\mathbb{R}^3$) are necessary for the vector $[a, b]$ to be perpendicular to a and b, i.e., to ensure that $(a, [a, b]) = (b, [a, b]) = 0$. One may verify that with $a^2 = (a, a)$ the standard formulas

$$[a, b]^2 = a^2 b^2 - (a, b)^2 , \tag{9.2}$$

and

$$([a, b], c)^2 = a^2 b^2 c^2 - a^2 (b, c)^2 - b^2 (a, c)^2 - c^2 (a, b)^2 + 2(a, b)(b, c)(c, a) \tag{9.3}$$

in $\mathbb{R}^3$ remain valid in $\mathbb{M}^3$.

Let $a, b \in \mathbb{M}^3, a \ll b$ and $p, q \in \beta C_a^+ \cap \beta C_b^-$ (Fig. 9.1). Introduce the four light ray segments

$$\begin{aligned}
l_1 &= p - a , \\
l_2 &= q - a , \\
l_3 &= b - q , \\
l_4 &= b - q .
\end{aligned} \tag{9.4}$$

We are now ready to prove:

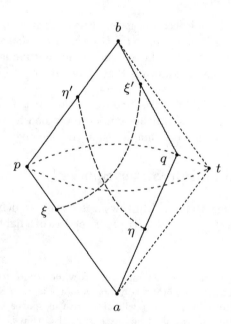

Fig. 9.1. The cushion problem

Theorem 9.2.1 (There are no cushions in Minkowski 3-space)

The four points a, b, p, q shown in Fig. 9.1 lie on a hyperboloid of one sheet.

Proof: We have to show that there exist suitable $o \in \mathbb{M}^3$ and $m \in \mathbb{R}$ such that the four equations

$$(\{a, b, p, q\} - o)^2 = -m^2 \tag{9.5}$$

are satisfied simultaneously.[2] Using (9.4), equations (9.5) may be written as

$$\begin{aligned}
(a - o)^2 &= -m^2 \,, \\
(a - o + l_1 + l_4)^2 &= -m^2 \,, \\
(a - o + l_1)^2 &= -m^2 \,, \\
(a - o + l_2)^2 &= -m^2 \,.
\end{aligned} \tag{9.6}$$

We set

$$a - o = c$$

and rewrite (9.6) as

$$\begin{aligned}
-c^2 &= m^2 \,, \\
(c, l_1) + (c, l_4) + (l_1, l_4) &= 0 \,, \\
(c, l_1) &= 0 \,, \\
(c, l_2) &= 0 \,.
\end{aligned} \tag{9.7}$$

Observe that c is perpendicular to l_1 and l_2. By (9.1), this implies that c has the form

$$c = \xi[l_1, l_2] \,. \tag{9.8}$$

From the second equation of (9.6) we obtain

$$\xi = -\frac{(l_1, l_4)}{([l_1, l_2], l_4)} \,. \tag{9.9}$$

and

$$m^2 = -\frac{(l_1, l_4)^2 [l_1, l_2]^2}{([l_1, l_2], l_4)^2} \,. \tag{9.10}$$

Using (9.2) and (9.3) this can be rewritten as

$$m^2 = \frac{(l_1, l_4)^2 (l_1, l_2)^2}{2(l_1, l_2)(l_2, l_4)(l_4, l_1)} = \frac{(l_1, l_4)(l_1, l_2)}{2(l_2, l_4)} \,. \tag{9.11}$$

The existence of $o = a - c$ and m^2 establishes the required result.   ■

---

[2] Here and in the following, $\{W, X, Y, Z\}$ will denote an ordered four-element set of which only one element is chosen at a time, as in (9.5).

Proposition 9.2.2 *Let $a, b \in \mathbb{M}^3$ be on the hyperboloid $x^2 = -m^2$. If $a - b$ is lightlike then the entire ray $l_{a,b}$ lies on the hyperboloid $x^2 = -m^2$.*

Proof: The ray $l_{a,b}$ consists of the points $b + \mu(a - b), \mu \in \mathbb{R}$. Now

$$
\begin{aligned}
(b + \lambda(a - b))^2 &= b^2 + 2\mu(b, a - b) + (a - b)^2 \\
&= b^2 + 2\mu(b, a - b),
\end{aligned}
\tag{9.12}
$$

since $(a - b)^2 = 0$. Next, $(a - b)^2 = (a - b)(a - b) = 0$ implies that $(a, a - b) = (b, a - b)$, and therefore $2(b, a - b) = a^2 - b^2$. But $a^2 = b^2(= -m^2)$, and therefore $a^2 - b^2 = 0$. Therefore

$$
(b + \mu(a - b))^2 = b^2 = -m^2
$$

for every μ, i.e., every point on $l_{a,b}$ lies on the hyperboloid $x^2 = -m^2$. Since in the three-dimensional Minkowski space the dimension of the hyperboloid is two, the two families of light rays in a light ray quadrilateral have to cross pairwise. This establishes the theorem. ∎

Theorem 9.2.1 leads quickly to the general result:

Theorem 9.2.3 (There are no cushions in Minkowski spaces)

Let $a, b, p, q \in \mathbb{M}^n$, $n > 3$ in the configuration shown in Fig. 9.1, i.e., a, b, p are coplanar but q lies outside this plane. Then a, b, p, q lie on a two-dimensional hyperboloid of one sheet.

Proof: Let $V \subset \mathbb{M}^n$ be the affine (or linear) subspace of $\mathbb{M}^n$ spanned by the light ray segments $l[a, p]$, $l[a, q]$ and $l[b, p]$. Then $V = \mathbb{R}^3$. The proof of Theorem 9.2.1 is coordinate-free, and therefore applies without change to the present situation. Note that the point c determined by (9.7) lies in V. ∎

The result that Minkowski spaces are cushion-free is easily transcribed to de Sitter spaces.

Theorem 9.2.4 *There are no cushions in de Sitter spaces.*

Proof: The de Sitter space $\mathbb{V}^n$ of dimension n can be embedded in $(n + 1)$-dimensional Minkowski space $\mathbb{M}^{n+1}$ such that

$$
\mathbb{V}^n = \{x | x \in \mathbb{M}^{n+1}, x^2 = -m^2, m > 0\} .
$$

In this embedding light rays of the de Sitter space coincide with light rays of the ambient Minkowski space. Therefore all relations between light rays in Minkowski spaces remain true in de Sitter spaces. ∎

9.3 Timelike Curves in Cushion-Free Spaces

In this section we shall assume that there are no cushions in M. Under these circumstances, the construction of timelike curves can be greatly simplified, as pairs of light ray segments $l[\xi, \xi']$ and $l[\eta, \eta']$ that are not known to intersect in the general case (Fig. 9.1) do intersect in Minkowski spaces. Figures 8.1 (p. 104), 8.2 (p. 106) and 8.3 (p. 107) illustrate the complications of the general case.

Let U be a D-set, $a, b \in U$, and $a \ll b$. As before, we shall construct a timelike segment $\Theta[a, b]$ that connects a with b. The construction given below will be "natural", although it will depend on the choice of a light ray segment.

Consider the order interval $I[a, b] \subset U$. Fix, arbitrarily, three points $p, q, r \in S(a, b)$ (see Fig. 9.2) and consider the ray segments

$$l[a, p], l[a, q], l[r, b] \text{ and } l[q, b] .$$

For brevity, we rename these segments as follows:

$$
\begin{aligned}
l[a, p] &= l[1] \\
l[a, q] &= l[2] \\
l[r, b] &= l[3] \\
l[q, b] &= l[4] .
\end{aligned}
\tag{9.13}
$$

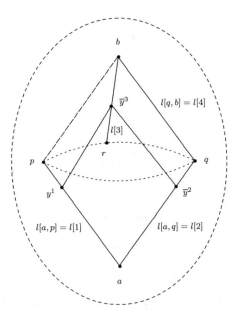

Fig. 9.2. Coordinates on $l[a, p]$ and $l[a, q]$

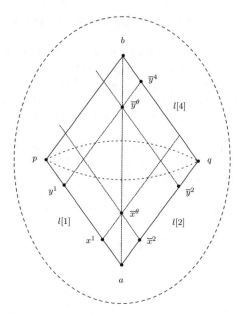

Fig. 9.3. Construction of timelike curves in cushion-free spaces

We specialize the maps ρ and σ defined by (5.1) and (5.2) to the above segments as follows:

$$\rho_{13} : l[1] \rightarrow l[3]$$
$$\sigma_{32} : l[3] \rightarrow l[2] \qquad (9.14)$$
$$\rho_{14} : l[1] \rightarrow l[4] \ .$$

Let y^1 be a point on $l[1]$; define

$$\overline{y}^3 = \rho_{13}(y^1)$$
$$\overline{y}^2 = \sigma_{32}(\overline{y}^3) \qquad (9.15)$$
$$\overline{y}^4 = \rho_{14}(y^1) \ .$$

What we have done is to take the point y^1 on $l[1]$, and map it to $l[2]$ and $l[4]$ by natural maps mediated by light rays. In the above, a bar over the letter indicates that the point is the image under a map ρ or σ and depends on the choice of the segment $l[3]$.

We now define a point $\overline{y}^\theta \in I[a, b]$ as follows[3] (see Fig. 9.3):[4]

---

[3] Again, the bar over y in $\overline{y}^\theta$ indicates that the point depends on the choice of the segment $l[3]$.

[4] In the figure, points of intersection of light rays are marked by dots. We adhere to the convention that ray crossings that are not so marked need not be points of intersection.

$$\overline{y}^\theta = l[y^1, \overline{y}^4] \cap \beta C_{\overline{y}^2}^+ . \tag{9.16}$$

Now let x^1 and y^1 be two distinct points on $l[1]$, and $\overline{x}^2, \overline{y}^2$ their images on $l[2]$ under the composite map $\sigma_{32} \circ \rho_{13}$. Let $\overline{x}^\theta, \overline{y}^\theta$ be the points of $I(a, b)$ determined by them by eq. (9.16). Then

$$x^1 <^l y^1 \Rightarrow \overline{x}^2 <^l \overline{y}^2 . \tag{9.17}$$

Furthermore, we can show that

$$x^1 <^l y^1 \Rightarrow \overline{x}^\theta \ll \overline{y}^\theta , \tag{9.18}$$

and the same with order reversed. We therefore define the segment $\Theta[a, b]$ by

$$\Theta[a, b] = \{\overline{y}^\theta \,|\, \overline{y}^\theta = l[y^1, \overline{y}^4] \cap \beta C_{\overline{y}^2}^+, \; y^1 \in l[1]\}. \tag{9.19}$$

The proof that $\Theta[a, b]$ is a timelike curve is fairly straightforward, and is omitted.

It should perhaps be remarked that, in a D-set in which light ray segments are differentiable, naturally constructed timelike curves are also likely to be differentiable.

If M is cushion-free, then the proof of (9.18) is routine. *We have not succeeded in proving that* (9.18) *holds if M is not cushion-free.*

10

Related Works

The special and general theories of relativity have led to many attempts at axiomatization of the theories, as well as of the notion of space-time. Some of it appears to have been explicitly motivated by (see, for example, [44]) Hilbert's sixth problem: "Mathematical Treatment of the Axioms of Physics" [15]. The work presented in this monograph is, by contrast, much more limited in scope. The quotations given below will set this in perspective.

This is how Hilbert formulated his sixth problem:[1]

> The investigations on the foundations of geometry suggest the problem: *To treat in the same manner, by means of axioms, those physical sciences in which mathematics plays an important part; in the first rank are the theory of probabilities and mechanics.*

And this is how he described his work on the foundations of geometry:

> The following investigation is a new attempt to choose for geometry a *simple* and *complete* set of *independent axioms* and to deduce from these the most important geometrical theorems in such a manner as to bring out as clearly as possible the significance of the different groups of axioms and the scope of the conclusions to be derived from the individual axioms [49].

By contrast with Hilbert's axiomatization of geometry or his sixth problem, our axiomatization is that of a single physical principle. Moreover, we have confined our investigations to determining the further *mathematical* structures that this single physical principle imposes (or suggests) on the point set on which it is defined. There is a considerable distance between our investigations and the axiomatization of a physical theory that has to satisfy the causality principle. The plain fact is that further hypotheses are required

---

[1] We may refer the reader who is interested in Hilbert's sixth problem to the article by A. S. Wightman [127].

H.-J. Borchers and R.N. Sen: *Mathematical Implications of Einstein–Weyl Causality*,
Lect. Notes Phys. **709**, 137–146 (2006)
DOI 10.1007/3-540-37681-X_10   © Springer-Verlag Berlin Heidelberg 2006

to arrive at even something so basic as the space-time of special relativity, as was indicated in Sect. 8.4.

There is, therefore, little in common between our investigations and the axiomatization of relativity theories or of the concepts of space-time. There are, however, two works that are exceptions to this rule. These are the works by Kronheimer and Penrose [66] and by Ehlers, Pirani and Schild [32].

Kronheimer and Penrose (hereafter KP) set up a scheme that is manifestly more general than ours for the study of global features of causal spaces, but not detailed enough to study their local structure as we have done. The abstract of their paper reads as follows:

> The paper examines the structure obtained by abstracting from the conventional (manifold) representation of relativistic space-time the concept of an event-set equipped with two partial orderings, whose counterparts are the notions "causally precedes" and "chronologically precedes in the history of some observer".

However, they make "no attempt to introduce sufficient axioms to reproduce all relevant properties of that model" [space-time], but rather "keep the axiom system as small and physically reasonable as possible", with the following aims in view (we quote from their paper):

1. To analyse unusual or pathological features of space-time manifolds for which no criterion of physical admissibility may otherwise be evident.
2. To admit structures which can be very different from a manifold [such as discrete space-times].

Ehlers, Pirani and Schild [32] address a different problem, and carry out a strictly local analysis. Starting from Weyl's observation [116] that a space-time manifold carries a host of different mathematical structures – topological, differential, conformal, projective, affine and (pseudo)-metric – that are related to each other, they ask which of these structures should be considered basic and which derived. They state their aim as follows:

> We wish to show how the full space-time geometry [conformal, projective, affine and pseudo-Riemannian] can be synthesized from a few assumptions about light propagation and free fall.

Chief among their assumptions is sufficient differentiability for the paths of light rays and freely falling particles. As our analysis is devoted to elucidating the local structure of the space, firstly without assuming differentiability, and then embedding it (densely) in one on which *differentiability may be assumed*, it falls in the no-man's land between the works of Kronheimer and Penrose, on the one hand, and of Ehlers, Pirani and Schild, on the other.

We shall describe parts of these works in sufficient detail to justify the above statement. Finally, we shall give some references to works on the axiomatization of Minkowski space and of relativity theory that we have come across. We are not specialists in these areas, and apologize in advance to authors whose works we may have overlooked.

Acknowledgment: The authors were not aware of the work of Kronheimer and Penrose [66] when [10] and [11] were written. They would like to thank H. Goenner for drawing their attention to it.

10.1 The Work of Kronheimer and Penrose

Recall that the aim of the authors was to provide a minimal axiomatization of the notion of causality, so that the resulting class of causal spaces would be as large as possible. We begin with a brief description of the axioms and some key definitions, *adhering to the original notations*,[2] but changing the order of presentation to suit our purposes.

KP 1.2: Definition

A *causal space* is defined to be a quadruple $\{X, \prec, \ll, \rightarrow\}$ where X is a set, $\prec$ and $\ll$ are two partial orders and $\rightarrow$ a relation defined on X. They are required to satisfy the following conditions ($x, y, z \in X$):

1. $x \prec x$.
2. If $x \prec y$ and $y \prec z$ then $x \prec z$.
3. If $x \prec y$ and $y \prec x$ then $x = y$.
4. $x \not\ll x$.
5. If $x \ll y$ then $x \prec y$.
6. If $x \prec y$ and $y \ll z$ then $x \ll z$.
7. If $x \ll y$ and $y \prec z$ then $x \ll z$.
8. $x \rightarrow y$ if and only if $x \prec y$ and $x \not\ll y$.

Each point a of a causal space X has a *causal future*, a *chronological future* and a *null future* which are defined respectively by the sets

$$J^+(a) = \{x | a \prec x\} \,,$$
$$I^+(a) = \{x | a \ll x\} \,, \tag{10.1}$$
$$C^+(a) = \{x | a \rightarrow x\} = J^+(a) \setminus I^+(a) \,.$$

The past sets $J^-(a), I^-(a)$ and $C^-(a)$ are defined similarly, by reversing the order.

KP 1.3: The *Alexandrov topology* $\mathcal{T}^*$ on a set X equipped with the partial order $\ll$ is defined as the coarsest topology on X in which each set $I^\pm(x)$ is open.

---

[2] A number such as **KP 1.2:** means that the statement that follows is taken from in Sect. 1.2 of KP, i.e., of [66].

10.1.1 The Relation $\rightarrow$

KP 1.1: Definition A reflexive relation R will be called *horismotic* if, for every finite sequence $\{x_i | x_i \in X, 1 \le i \le n - 1\}$ such that $x_i R x_{i+1}$, and all positive integers h, k satisfying $1 \le h \le k \le n$,

1. $x_1 R x_n \Rightarrow x_1 R x_k$, and
2. $x_n R x_1 \Rightarrow x_h = x_k$.

Any reflexive partial ordering is horismotic.

KP Lemma 1-1 *Let x, y, z be points in a causal space. If $x \prec y \prec z$ and $x \rightarrow z$, then $x \rightarrow y \rightarrow z$.*

KP 2.1: Construction of a causal space from the horismos

Let X be a set and $\rightarrow$ a horismotic relation on it. Define the relations $\prec^{\mathfrak{A}}$ and $\ll^{\mathfrak{A}}$ on it as follows:

1. $x \prec^{\mathfrak{A}} y$ *iff there exists a finite sequence $\{u_i | u_i \in X, 1 \le i \le n\}$ which satisfies*

$$x = u_1 \rightarrow u_2 \rightarrow \cdots \rightarrow u_n = y; \qquad (\text{KP } 2 \cdot 1 \cdot 1)$$

2. $x \ll^{\mathfrak{A}} y$ *if $x \prec^{\mathfrak{A}} y$ and $x \nrightarrow y$.*

Then $\{X, \prec^{\mathfrak{A}}, \ll^{\mathfrak{A}}, \rightarrow\}$ is a causal space.

10.1.2 Comparison with the Present Work

The following identifications are obvious, and unique:

1. The order $\prec$ of KP may be identified with our $<$, i.e., $x \prec y \equiv y \in C_x^+$. Then condition 3) of def. KP 1.2 becomes our cone axiom.
2. The order $\ll$ of KP may be identified with our $\ll$. Then conditions 4)–7) of def. KP 1.2 will be satisfied.
3. With these identifications, the relation $x \rightarrow y$ of KP will have the following meaning: $x \rightarrow y \equiv y \in \beta C_x^+$. However, in our scheme $y \in \beta C_x^+ \equiv x <^l y$, so that $x \rightarrow y \equiv x <^l y$.

The rendering of (KP2 $\cdot$ 1 $\cdot$ 1) into our scheme will read as follows:

$x < y$ iff there is an ascending l-polygon from x to y.

The rendering of KP lemma 1-1 into our scheme will read as follows:

Let x, y, z be points in a causal space. If $x < y < z$ and $x <^l z$, then $x <^l y$ and $y <^l z$.

If now x, y, z lie in a D-set then, in our scheme, the convexity axiom[3] would require x, y, z to lie on the same light ray. However, *the KP axioms do not make this demand*, which is precisely what makes the relation $\rightarrow$ nontrivial in KP. Lemma KP 1-1 means that the light ray segments $l[x, y]$, $l[y, z]$ and $l[x, z]$ form a triangle which lies on βC_x^+, which is impossible in Minkowski spaces.

In our scheme, the crucial Theorem 5.1.3, which states that the standard maps ρ and σ (defined by (5.1) and (5.2) respectively) are bijective and are the inverses of each other, will not hold without the convexity axiom, as shown by the example that follows. This in turn will entail the failure of the entire development from Chap. 5 onwards, that is of the entire theory of the local structure of ordered spaces as developed in the present work.

The example that follows is established under a condition that is weaker than the convexity axiom. It is:[4]

If $x \in U$ and $l \cap \beta C_x^+$ contains two distinct points, then

$$l \cap \beta C_x^+ \cap U \subset \beta C_x^+ ,$$

and the same for reversed order.

Example 10.1.1 Let $M = \mathbb{R}^3$ and $x = (x_0, x_1, x_2)$. Define the cones $C_x^{\pm}$ at x as follows:

$$C_x^+ = \{z | z \in \mathbb{R}^3, (z_0 - x_0)^2 - (z_1 - x_1)^2 - (z_2 - x_2)^2 \geq 0 ,$$
$$z_0 - x_0 \geq 0, z_1 - x_1 \leq 0\} ,$$

$$C_x^- = \{z | z \in \mathbb{R}^3, (z_0 - x_0)^2 - (z_1 - x_1)^2 - (z_2 - x_2)^2 \geq 0,$$
$$z_0 - x_0 \leq 0, z_1 - x_1 \geq 0\} .$$

The boundaries of these cones will consist of two parts, the "normal" part which we shall denote by $\beta C_x^{\pm}(N)$ and the "vertical" part which we shall denote by $\beta C_x^{\pm}(V)$. Explicitly,

$$\beta C_x^+(N) = \{z | z \in \mathbb{R}^3, (z_0 - x_0)^2 - (z_1 - x_1)^2 - (z_2 - x_2)^2 = 0 ,$$
$$z_0 - x_0 \geq 0, z_1 - x_1 \leq 0\}$$

$$\beta C_x^+(V) = \{z | z \in \mathbb{R}^3, z_0 - x_0 \geq 0, z_1 - x_1 = 0 ,$$
$$(z_0 - x_0)^2 - (z_2 - x_2)^2 \geq 0\},$$

$$(10.2)$$

---

[3] Recall that the convexity axiom is the name given to condition d) of the defining Properties 4.2.1 of D-sets.

[4] This is condition d) of the definition II.2.1 of D-sets in [10], and is more or less the same as lemma KP 1-1. The Example 10.1.1 given above, found after [10] was published, showed that this condition was too weak to enforce the desired results. It was therefore replaced by condition d) of 4.2.1 in [11].

with

$$\beta C_x^+ = \beta C_x^+(N) \cup \beta C_x^+(V) .$$

The parts $\beta C_x^-(N)$ and $\beta C_x^-(V)$ of βC_x^- may be written down analogously. The light rays through x are the straight lines through x lying on βC_x.

Choose now two points

$$x = (0,0,0) \text{ and } y = (2,-1,0) \tag{10.3}$$

and two light rays

$$l_x = \{a(1,0,-1)|a \in \mathbb{R}\}$$
$$l_y = \{(2,-1,0) + b(-1,1,0)|b \in \mathbb{R}\} \tag{10.4}$$
$$= \{(2-b, b-1, 0)|b \in \mathbb{R}\} .$$

Clearly, $y \gg x$. One may verify that $l_x^+ \subset \beta C_x^+(N)$ and $l_y^- \subset \beta C_y^-(N)$.

Next, we define the points $p = l_x^+ \cap \beta C_y^-$ and $q = l_y^- \cap \beta C_x^+$. A short calculations shows that

$$p = \left(\frac{3}{4}, 0, -\frac{3}{4}\right) ,$$
$$q = (1,0,0) . \tag{10.5}$$

It follows from the above that

$$l_x[x,p] = \left\{(a,0,-a)\Big|0 \le a \le \frac{3}{4}\right\}$$
$$l_y[q,y] = \{(2-b, b-1, 0)|0 \le b \le 1\} . \tag{10.6}$$

We want to determine, explicitly, the map $\rho : l_x[x,p] \to l_y[q,y]$ defined by $\rho(r) = \beta C_r^+ \cap l_y[q,y]$, where $r \in l_x[x,p]$.

Substituting $r = (a,0,-a)$ for x in (10.2), we obtain the conditions that determine $\beta C_r^+(N)$ and $\beta C_r^+(V)$ in terms of a. Setting $z = (2-b, b-1, 0)$ into these conditions, we obtain the conditions on a and b that determine the intersections $l_y \cap \beta C_r^+(N)$ and $l_y \cap \beta C_r^+(V)$. Carrying out the calculations, we find that

1. $l_y \cap \beta C_r^+(N) \ne \emptyset$ iff

$$b = \frac{3-4a}{2(1-a)} \tag{10.7}$$

or, equivalently

$$a = \frac{3-2b}{2(2-b)} . \tag{10.8}$$

2. $l_y^- \cap \beta C_r^+(V) \neq \emptyset$ iff

$$0 \le a \le \frac{1}{2}, \tag{10.9}$$

and for every value of a in this range,

$$b = 1.$$

Equations (10.7)–(10.9) determine the behaviour of the map $\rho : l_x[x, p] \rightarrow l_y[q, y]$. We find that:

1. The segment $l_x[(\frac{1}{2}, 0, -\frac{1}{2}), (\frac{3}{4}, 0, -\frac{3}{4})]$ is mapped one-to-one onto the segment $l_y[(1, 0, 0), (2, -1, 0)]$.

2. The segment $l_x[(0, 0, 0), (\frac{1}{2}, 0, -\frac{1}{2})]$ is mapped into a single point $q = (1, 0, 0)$ on $l_y[q, y]$.

By contrast with KP, we start from light rays and the order $<^l$ on them. In the language of KP, we abstract from the event set only the causal and not the chronological order. We define the order relations $<$ and $\ll$ (which are parallel to the $\prec$ and $\ll$ of KP) in terms of $<^l$. The main differences between our ansatz and that of KP are:

1. Our spaces are dense subsets of complete spaces. They allow totally disconnected spaces such as $\mathbb{Q}^2$ and its higher-dimensional analogues $\mathbb{F}^n$ (see Example 5.2.2, and also footnote 3 on p. 58).[5] The density axiom on light rays (part b of Axiom 3.1.2) has the big advantage that τ-interiors and β-boundaries of light cones can be defined without the topology.

2. In the language of topological vector spaces (when appropriate), our cones have one-dimensional faces. KP allow general faces.

Our definition of D-sets is none other than a basis for an improved version of the Alexandrov topology.

10.2 The Work of Ehlers, Pirani and Schild

The background to the work [32] of Ehlers, Pirani and Schild is a certain problem on the foundations of general relativity, on which there exists a very substantial literature. However, this problem lies well beyond our scope. For a brief but lucid summary, we refer the interested reader to a later work by Woodhouse [131]. Further references may be found in this work and in the bibliography of [32].

---

[5] In the physics literature, one often finds the term *discrete space* used where the term *totally disconnected* may well be more appropriate, as in the phrase "[such as discrete space-times]" in one of the quotations from KP on p. 138. In mathematical usage, a discrete (topological) space is one in which one-point sets are open, and this topology can have no relation to any causal structure.

10.2.1 Notations and Terminology

The terms *event, particle, message* and *echo* are often used in the literature on relativity theory. *Event* denotes a point in space-time, which is generally assumed to be a 4-manifold. *Particle* denotes the space-time trajectory or *world line* of a physical point-particle with mass. Such a trajectory is ordinarily a timelike curve.

Let P, Q be two particles, p a point on P and $q = \beta C_p^+ \cap Q$. It is assumed that the intersection is nonempty. The map defined by $m : p \mapsto q$ is called a *message* from P to Q. Now let $p' = \beta C_q^+ \cap P$. The formula $p' = e(p)$ defines a map $e : P \to P$ that is called the *echo* on P from Q. Often the domains and co-domains of these maps are not specified any further.

Let e be an event and P, P' two distinct particles. Let

$$
\begin{aligned}
u &= \beta C_e^- \cap P\,, \\
r &= \beta C_e^+ \cap P\,, \\
u' &= \beta C_e^- \cap P'\,, \\
r' &= \beta C_e^+ \cap P'\,.
\end{aligned}
\tag{10.10}
$$

If P and P' are each furnished with a real parameter (local coordinates), then (10.10) defines a map $x_{PP'} : e \mapsto (u, u', r, r')$ from M to $\mathbb{R}^4$. The numbers u, u', r, r' are called the *radar coordinates* of e.

EPS use M to denote the underlying (4-dimensional) space and $\mathcal{C}, \mathcal{P}$ and $\mathcal{A}$ to denote, respectively, the conformal, projective and affine structures on it. For $p \in M$, they denote the tangent space at p by M_p and the projective 3-space of directions at p by D_p. They use L_1, L_2 to denote light rays, and L_e to denote the set of lightlike directions at the event e.

10.2.2 The EPS Axioms

Axiom D_1: *Every particle is a smooth, one-dimensional manifold; for any pair P, Q of particles, any echo on P from Q is smooth and smoothly invertible.*

Axiom D_2: *Any message from a particle P to another particle Q is smooth.*

Axiom D_3: *There exists a collection of triplets (U, P, P'), where $U \subset M$, $P, P' \in \mathcal{P}$ such that the system of maps $x_{PP'}|_U$ is a smooth atlas for M. Every other map $x_{QQ'}$ is smoothly related to the local coordinate system of that atlas.*[6]

Axiom D_4: *Every light ray is a smooth curve in M. If $m : p \mapsto q$ is a message from P to Q, then the initial direction of L at p depends smoothly on p along P.*

---

[6] Radar coordinates cannot be defined on odd spacetime dimensions. However, alternatives that apply to odd as well as even dimensions are easily devised.

Axiom L_1: *Any event e has a neighbourhood V such that each event p in V can be connected within V to a particle P by at most two light rays. Moreover, given such a neighbourhood and a particle P through e, there is another neighbourhood $U \subset V$ such that any event p in U can, in fact, be connected with P within V by precisely two light rays L_1, L_2, and these intersect P in two events e_1, e_2 if $p \notin P$. If t is a coordinate on $P \cap V$ with $t(e) = 0$, then $g : p \mapsto -t(e_1)t(e_2)$ is a function of class C^2 on U (see Fig. 10.1).*

Axiom L_2: *The set L_e of light-directions at an (arbitrary) event e separates $D_e - L_e$ into two connected components. In M_e the set of all non-vanishing vectors that are tangent to light rays consists of two connected components.*

Axiom P_1: *Given an event e and a C-time-like direction D at e, then there exists one and only one particle P passing through e with direction D.*

Axiom P_2: *For each event $e \in M$, there exists a coordinate system $(\bar{x}^a)$, defined in a neighbourhood of e and permitted by the differential structure introduced in Axiom D_3, such that any particle P through e has a parameter representation $\bar{x}^a(\bar{u})$ with*

$$\left. \frac{d^2 \bar{x}^a}{d\bar{u}^2} \right|_e = 0 ; \tag{10.11}$$

such a coordinate system is said to be projective at e.

Axiom C: *Each event e has a neighbourhood U such that an event $p \in U, p \neq e$ lies on a particle P through e if and only if p is contained in the interior of the light cone ν_e of e.*

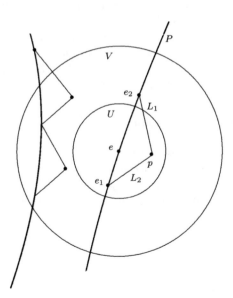

Fig. 10.1. Illustrating Axiom L_1

Clearly, EPS are assuming that M has a differentiable structure, and that the order structure itself is differentiable.

If the cushion problem is answered in the negative, then one would have at hand a class of timelike curves that are defined naturally, and one could expect the differentiability of light rays to extend to these timelike curves as well. This would enable a fairly smooth transition to be established between our work and EPS.

To conclude this section, we refer to the paper by Woodhouse [131] that was mentioned earlier. This paper gives a somewhat different version of the work of Ehlers, Pirani and Schild.

10.3 Other Works

The works mentioned in this brief section are works that we have come across. They are rather far from our areas of specialization and we are not in a position to do justice to them. For the same reason, it is entirely possible that there are other works of similar scope and significance that we have not come across; omission from this section should be interpreted as a reflection of our unfamiliarity with the field, and not as a judgment.

10.3.1 The Monograph of Schutz

Schutz [95] has a monograph entitled *Independent axioms for Minkowski space-time*, published in 1997. This volume has an introductory chapter that covers i) axiomatic systems, ii) independence and consistency of the set of axioms, iii) axiomatic systems for geometries, iv) axiomatic systems for space-times and v) a brief introduction to the [author's] present axiomatic system. The rest of the books develops this theme. It has, however, some appendices of wider interest, and a fairly extensive bibliography.

10.3.2 Works of Soviet Scholars

Soviet scholars have followed the lead of A. D. Alexandrov and developed a school on the foundations of relativity theory or axiomatic relativity. We may cite the review by Guts [44], entitled *Axiomatic Relativity Theory* (which has an extensive bibliography), and a slightly later work by Guts and Levichev [45], which has some further results.

11

Concluding Remarks

The remarks in this chapter are divided into three parts: Concerning physics, concerning mathematics and concerning Cantor, Wigner and Popper.

11.1 Concerning Physics

The theory developed in these pages may give the impression that we are dealing mainly with a branch of mathematics. However, our endeavour is based solidly on ideas prevailing in contemporary physics. Therefore some remarks on the physical background and the problems raised by it that are currently under investigation may not be out of place.

The backbone of natural sciences is the fact that experiments can be repeated at different times and at different places, and that the outcomes of the same experiment performed at different times and places turn out to be the same, within experimental errors. The 'scientific method' assumes that a background of space and time or space-time is given, and that one knows how to compare experiments at different places and different times. A deeper analysis of this assumption leads to the concept of symmetry (see, e.g., Ekstein [36, 37] and Avishai and Ekstein [5]). The notion of indefinite repeatability of experiments may be considered as validated for laboratory physics, but it is not relevant to astronomy, where one does not have the possibility of controlling parameters that one has in the laboratory.[1] One observes events that may be widely separated in space and time – for example, γ-ray bursts – and one needs a theory that can compare events that are so separated. If the background is given, then one can speak about the position of a material object. In quantum mechanics the determination of position is restricted by the uncertainty principle. In classical physics, one may say that it is restricted by the Gaussian theory of errors. In both cases a point is effectively replaced by

---

[1] Indeed, astronomers call their subject an *observational* and not an experimental science.

H.-J. Borchers and R.N. Sen: *Mathematical Implications of Einstein–Weyl Causality*,
Lect. Notes Phys. **709**, 147–156 (2006)
DOI 10.1007/3-540-37681-X_11

a (continuous) distribution function. The C^*-algebra formed by these distributions reconstructs the background space via the Gelfand isomorphism (see [24, 81]).

If space-time itself becomes an object of scientific investigation, the situation becomes much more complicated. Since two or more experiments cannot generally be performed simultaneously at the same place, the assumption that the physical space-time continuum consists of points becomes a theoretical construct which is not directly testable by experiment. The postulate of a continuum of events can only be justified if the testable predictions of a theory based on this postulate are verified experimentally.[2] This is indeed the case for Einstein's general theory of relativity [34]. One local example is the precession of the perihelion of Mercury. In the 19th century this effect could not be fully explained by the influence of the other planets. The discrepancy that remained was explained by Einstein.

Another example that supports Einstein's general theory of relativity is the bending of light by the gravitational field of the sun. In 1911, Einstein calculated this effect by combining Newtonian gravitation with special relativity. He used the fact that in a homogeneous gravitational field all bodies fall equally fast, plus the equivalence of mass and energy.[3]

Let the energy of a system be E_0 at the point S_0. If it is raised by the height h to the point S_1, its energy will become $E_1 = E_0(1 + \gamma h/c^2)$. If S_0 may be regarded as the position of a gravitating mass-point, this formula may be written as $E_1 = E_0(1 + \Phi/c^2)$, where Φ is the Newtonian potential. From this one obtains the formula

$$c = c_0 \left(1 + \frac{\Phi}{c_0^2} \right) , \tag{11.1}$$

for the velocity c of light at S_1. With this Einstein derived, as in geometrical optics, the general formula

$$\alpha = \frac{1}{c^2} \int \frac{\partial \Phi}{\partial n'} ds \tag{11.2}$$

for the bending of a light ray. In the above, ds is the line element and n' is the normal to the gradient of the potential. When the gravitational field is that of a single star, (11.2) becomes

$$\alpha = \frac{2kM}{c^2 \Delta} , \tag{11.3}$$

where m is the mass of the star, k is the gravitational constant and Δ is the distance from the centre of the star to the closest point on the light ray.

---

[2] At this point the reader may be referred to Popper's work [84], the brief discussion in Sect. 11.3, and the article [96].

[3] See the report by Pauli [80].

Assuming the star to be the sun, Einstein obtained the figure $\alpha = 0''.83$. However, the general theory of relativity, in which light rays were null-geodesics, predicted a displacement $\alpha = 1''.75$. Details of the calculation may be found in most textbooks.

The British expeditions of 1919 that observed this bending [31] during the total solar eclipse confirmed the general relativity value.[4] Just after the horrors of World War I, a British scientific expedition had overthrown Newton and confirmed the theory of an enemy scientist! It caught the imagination of the newspapers, and made Einstein's name into a household word.

Another local prediction of the general theory of relativity was that the change of the local metric[5] with time is negligible; this fact is put to use for the global positioning system (GPS). Hubble's discovery of the recession of galaxies in 1929 [51], made with the 100-inch Mt. Palomar telescope, started the modern era of cosmology. This, in turn, led to the model of the expanding universe. The observation of the $3°$ K cosmic microwave background radiation by Penzias and Wilson from the Bell Telephone Laboratories [83] reinforce the idea that the universe must have had a beginning; this beginning is usually called the *big bang*. Trust in the cosmological model is a matter of taste; not everyone likes the idea that most of the matter (dark matter) and the greater part of energy (dark energy) in the universe should be inaccessible to direct observation.

The standard theory is based on the concept of geometrical points, although one is aware of the fact that they are not accessible to experiment. Therefore some people believe that in the regime of Planck length ($\sim 10^{-33}$ cm) the concept of points should be replaced by something else. Doplicher, Fredenhagen, and Roberts [26] discussed this problem and argued that the coordinates x_k of a point is space should not commute, and invented commutation relations for the different components. Similar relations were used by Wess [111]. These have been other attempts to replace standard space-time by a non-commuting object, but they will not be discussed here. However, the construction of quantum field theory on such objects is still at a rudimentary state. (A quantum field theory on a manifold or on a generalized manifold corresponds to a theory of test particles which do not influence the structure of the manifold.) For an example, see [6].

There have been several attempts at a quantum theory of the gravitational field. One of them is string theory, on which we do not wish to comment. Another is the background-independent method of quantization developed by Ashtekar (see, e.g., [2]). But, as far as we know, there is as yet no theoretical replacement for the manifold, at least in ZFC-mathematics, in which the

---

[4] An account of later observations on quasars that improved the accuracy by an order of magnitude may be found in [90].

[5] Here and in the following the term *metric* is used in the sense of general relativity; see Sect. 2.2.1.

concept of points is not meaningful in the small. We have therefore based our investigation on the notion of sets of geometrical points.

Our second fundamental concept is that of light rays. Light rays are often exploited in everyday life for purposes other than seeing. In a factory, for instance, or for excavating tunnels, light rays are used to lay out straight lines. All such uses are based on the fact that the change of the metric with time is negligible and that flat space is a good approximation to reality. More generally, one should identify light rays with geodesics, as Gauß[6] did when he tried to determine the structure of space. (At that time space and time where independent entities.) He used large triangles with sides several kilometres long and measured the sum of the three angles; he could not detect any discrepancy from π. As vertices of the triangle Gauß used towers which were constructed for his triangulation of the palatinate of Hannover [41].[7]

Recall that we represented light rays mathematically by totally ordered sets which become endowed, upon completion, with one-dimensional differentiable structures.[8] According to H. Kneser [60] there exist, besides the circle, four other one-dimensional Hausdorff continua endowed with a $C^{(k)}$-structure. They are the real line, the two long half-lines, and the long line. Since length measurements determine distances between two points (which are compact intervals), and since one can make only a finite number of such measurements, a physical quantity of the nature of a length can be represented on the real

---

[6] In the text, we have used the modern German spelling of the name Gauß. In his own works, which were written in Latin, Gauß used the Latin spelling of his name. This spelling was retained in the German translation [41] cited in the bibliography, and we have adhered to the original.

[7] Details of the measurement are given in Sect. 28 of reference [41]. The largest triangle was between the mountains Brocken–Inselberg–Hoher Hagen, with the sides being approximately 105–85–70 km long. The difference of the sum of the three angles from $180°$ was less than $2''$.

[8] There appears to be little uniformity in the definition of a topological or differentiable manifold in the literature. The definition of a $C^{(k)}$-structure ($k = 0, 1, \ldots, \infty$) by means of compatible overlapping charts seems to be unique. But a space possessing a $C^{(k)}$-structure does not have to be Hausdorff, second countable or paracompact, and this is where definitions begin to differ. For example, Kneser [60] and Matsushima [75] require the Hausdorff condition but not second countability or paracompactness. Kosinski [64] requires a manifold to be Hausdorff and second countable, whereas Hirsch [50] requires one to be paracompact as well. Kobayashi and Nomizu, in their standard text (vol. 1 of [62]) introduce the assumption of paracompactness only on p. 58. The definition of a manifold that we used in Chap. 2 assumes all three conditions to ensure the existence of a Riemannian metric.

These distinctions are not academic. The line with two origins, for example, is second countable, paracompact but not Hausdorff; the long line is Hausdorff but not second countable or paracompact. Many other counterexamples may be found in [101].

line. This follows from the fact that a compact connected set on any of these four lines is isomorphic to a closed interval on the real line.

In Sect. 3.1 we stated that our mathematical notion of light rays corresponded to null geodesics (see footnote 2 on p. 16). This is true by definition in Minkowski space. In the general theory of relativity, Minkowski space is devoid of matter (and therefore of light), which leads to the unhappy conclusion that our definition of light rays can make sense only if there is no light! However, one does not have to put up with such misery. One can also abstract our mathematical notion of light rays from objects of geometrical optics, adapted to space-times of general relativity. Then the index of refraction may be interpreted as the effect of interstellar gases and dust, and a density gradient of the latter should lead to a deviation of light rays from geodesics. The assumption one has to make is that the refraction is *dispersion-free*. Recall that we also allowed two light rays to cross each other more than once (although not in a D-set; see Example 3.1.13). This happens not only in our theory (as a consequence of the global structure, as in the example quoted), but also in nature. This effect is called the formation of images by *gravitational lenses*, and is observed in nature as a consequence of the presence of large masses.

If the ordered space M is finite dimensional, then it can be embedded in a finite dimensional Euclidean space. In this situation one may look at the global structure of light rays. Not every light ray needs to tend to infinity at both ends. For instance a star produces holes in the manifold. These are timelike cylinders where the light rays might end or begin. If stars are born, then these cylinders have a cone-like beginning. If one wants to describe an expanding universe which starts with a big bang, then all light rays originate at a point or in some finite area, depending on the details of the model. For the description of black holes the situation is reversed in time, which means that some of the light rays end at black holes. Whether they end at the singularity or already at the horizon depends on the situation. Our axioms might be valid across the horizon and inside. If not the light rays will have to end at the horizon.

In Sect. 8.4 we showed that locally compact order complete spaces carry a local differentiable structure. This does not imply that there exists a global differentiable structure. The main obstruction is the fact that the structure is influenced by the choice of the timelike curve. The construction of the curve in Sect. 8.1 depends on the choice of the coordinates on $l[a,p]$ and $l[a,q]$ and the choice of the two points $p, q \in S(a,b)$ (see Fig. 8.1). If we change the points and the coordinates, then, in most cases, the two timelike curves will lead to different parametrizations of the D-interval which will not be diffeomorphic to each other. Therefore, when we have two D-intervals with non-empty intersection, the two parametrizations of the intersection will not be diffeomorphic to each other.

If one takes the viewpoint of a practitioner,[9] namely, that since it is impossible to distinguish between a continuous, nowhere differentiable and a $C^{(\infty)}$-manifold by a finite number of experiments, one might as well assume that one is dealing with a smooth space, then there remains another problem. We are dealing with an ordered space, and therefore we would like the interior of the cone mantle (without the vertex x) $\beta C_x^{\pm} \setminus \{x\}$ to be a $C^{(\infty)}$-submanifold of codimension one. This problem can only be solved if we assume isotropy, as we did in Sect. 8.4.3.

In Chap. 9 we discussed the cushion problem and saw that there are no cushions in Minkowski and de Sitter spaces. We do not know what this result tells us about the structure of a space that is known to be cushion-free. The universal covering of a de Sitter space can be mapped order equivalently onto $\mathbb{R}^n$ with the same dimension n, but with a non-Minkowski order. Therefore one might ask whether or not the universal covering of a cushion-free space can be mapped order equivalently into a Riemann surface over $\mathbb{R}^n$ of the same dimension.

Another problem that we have not discussed is the following: In Sect. 5.1 we introduced the standard homeomorphisms ρ and σ between light ray segments lying on the boundary of a D-interval $\beta I[x, y]$. Choosing now four different points $\{p_1, \ldots, p_4\} \in S(x, y)$ we can construct the chain

$$l[x, p_1] \longrightarrow l[p_2, y] \longrightarrow l[x, p_3] \longrightarrow l[p_4, y] \longrightarrow l[x, p_1]$$

and obtain a map from $l[x, p_1]$ onto $l[x, p_1]$. Changing the points $\{p_2, \ldots, p_4\}$ we obtain a group of maps of $l[x, p_1]$ onto itself which leaves the endpoint fixed. What is the structure of this group? If we change the D-interval, does then the structure of the group change?

Our investigation is concerned only with the conformal structure of physical space. If one wants to go on to the Weyl projective structure one would have to postulate that the two vertices of a D-interval are connected by a unique trajectory of a freely falling test particle. These curves would correspond to timelike geodesics. By changing the order interval we would obtain different geodesics. Such a trajectory between the vertices and be mapped onto the real interval $[-1, +1]$. Denoting the points on this curve by $p(t)$, we can define a spacelike hypersurface $\cup \{S(p(-t), p(+t)); 0 \leq t \leq 1\}$. There remains the problem of comparison of the scales for different D-intervals. This would be necessary for giving a meaning to the intersection of several spacelike hypersurfaces obtained from different trajectories containing a given point.

11.2 Concerning Mathematics

The notion of partial order, by itself, is probably too weak to be either interesting or useful; the category of "partially ordered spaces" will be much too

---

[9] A somewhat more skeptical view is discussed in Sect. 11.3.

large. The class that we have called ordered spaces is much more structured. "Forgetting" all the physics, one may define an ordered space as follows.

Definition 11.2.1 An *ordered space* is a point-set X with a nonreflexive and nonsymmetric partial order $\ll$ defined on it. This partial order is required to satisfy the following conditions (which are not necessarily independent):

1. $a, c \in X, a \ll c \Rightarrow \exists\, b \in X$ such that $a \ll b \ll c$.
2. If $x \in X$ then $\exists\, a, b \in X$ such that $a \ll x \ll b$.
3. For $a, b \in X, a \ll b$, define $I(a,b) = \{x | x \in X, a \ll x \ll b\}$. The collection $\mathcal{U} = \{I(a,b)\}$ covers X, and is a base for a topology $\mathcal{T}$ on X.
4. There exists a refinement $\mathcal{W} = \{I(\alpha, \beta)\}$ of $\mathcal{T}$ which is
   a) A cover for X.
   b) A base for the topology of X.
   c) Let $x \in I(a,b) \subset W$, where $W \in \mathcal{W}$. Then for points a', b' with $a \ll a' \ll b' \ll b$, one has $\bar{I}(a', b') \subset I(a,b)$. Here the bar above means topological closure.
   d) $I(\alpha, \beta), I(\alpha', \beta') \in \mathcal{W}$ implies that $I(\alpha, \beta)$ and $I(\alpha', \beta')$ are homeomorphic to each other.
   □

The ordered space defined above has the following property:

Theorem: *If $\{X, \mathcal{T}\}$ is Hausdorff, then it is a Tychonoff space.* ∎

Consequently, the pair $\{X, \ll\}$ is a mathematically useful, and therefore interesting, object.

Note that the Hausdorff assumption is essential; it does not follow from Def. 11.2.1. The topology of "Newton causality" described on p. 5 satisfies the conditions of Def. 11.2.1, but is not Hausdorff.

Furthermore, we have seen in Sect. 8.2.1 that an order complete space possesses certain separation properties analogous to the Jordan-Brouwer separation properties. An order structure imposed on $\mathbb{R}^{n+1}$ followed by the one-point compactification makes it easy to prove results analogous to the Jordan curve Theorem 8.2.12, because the notions of "inside" and "outside" are very precisely defined for order intervals. This may also be an avenue worth exploring.

Remark 11.2.2 Recall the definition of a symmetric relation R: $xRy \Rightarrow yRx$ *for all* x, y, so that the statement that R is nonsymmetric only means that there exist x, y such that $xRy \not\Rightarrow yRx$; the existence of pairs a, b such that $aRb \Rightarrow bRa$ is not excluded. However, if R is *transitive* – which is obligatory for a partial order – then aRb and bRa together imply aRa (and bRb), which contradicts the nonreflexivity of R. That is, if $\prec$ is a nonreflexive, nonsymmetric order relation, then $x \prec y$ necessarily implies that $x \neq y$.

The order relation $\ll$ defined in 11.2.1 has been abstracted from the relation $y \in \tau C_x^+$ (Theorem 3.2.22). The conditions that this relation is required to satisfy have been abstracted from the result that there exists a family of D-intervals that covers M and is a basis for a topology. Had we used the order $\prec$ abstracted from the relation $<$ (or $y \in C_x^+$) of Def. 3.2.6, the definition of the topology would have been more involved (as the sets $\{x | a \prec x \prec b\}$ would have been closed), and the transparency of Def. 11.2.1 would have had to be sacrificed.

11.3 Concerning Cantor, Wigner and Popper

It could be an understatement to say that Georg Cantor has not always been well served by the early writers who have written about the man and his work (see the Introduction to [20]). The following quotations are from the section *Nature of Mathematics*, pp. 132–133, in *Georg Cantor: His Mathematics and Philosophy of the Infinite*, by J. W. Dauben [20], a more recent work which was first published in 1979:

> Cantor asserted the reality of both the physical and ideal aspects of the number concept...It was one of the most difficult problems of metaphysics to determine the nature of the connection between the two.
>
> Cantor ascribed the necessary coincidence of these two aspects of number to the unity of the universe itself.[10] This meant that it was possible to study... [it] without having to confirm or conform to any objective content. This set mathematics apart from all other sciences and gave it an independence that was to imply great freedom for mathematicians in the creation of mathematical concepts. It was on these grounds that Cantor offered his now famous dictum that the essence of mathematics is its freedom. As he put it in the *Grundlagen* [17]:
>
>> Because of this extraordinary position which distinguishes mathematics from all other sciences, and which produces an explanation for the relatively free and easy way of pursuing it, it especially deserves the name of *free mathematics*, a designation which I, if I had the choice, would prefer to the now customary "pure" mathematics.[11]
>
> Mathematics, Cantor believed, was the one science justified in freeing itself from any metaphysical fetters. Applied mathematics and theoretical mechanics, on the contrary, were metaphysical in both their content and goals.[12]

---

[10] The author refers, at this point, to Cantor [17], p. 182.

[11] The author refers, at this point, to Cantor [17], p. 182.

[12] The author refers, at this point, to Cantor [17], p. 183.

Wigner, while agreeing with Cantor that the essence of mathematics was its freedom, was less convinced by "the unity of nature". In his much-quoted essay on *The unreasonable effectiveness of mathematics in the natural sciences* [128], he wrote:

> Somebody once said that philosophy is the misuse of a terminology which was invented just for this purpose.[13] In the same vein, I would say that mathematics is the science of skillful operations with concepts and rules invented just for this purpose. The principal emphasis is on the invention of concepts. Mathematics would soon run out of interesting theorems if these had to be formulated in terms of the concepts which already appear in the axioms. Furthermore, whereas *it is unquestionably true that the concepts of elementary mathematics and of elementary geometry were formulated to describe entities which are directly suggested by the actual world* [our emphasis], the same does not seem to be true of the more advanced concepts, in particular the concepts which play such an important role in physics. . .
>
> The physicist is interested in discovering the laws of inanimate nature. . .[14]
>
> Having refreshed our minds as to the essence of mathematics and physics, we should be in a better position to review the role of mathematics in physical theories. . .
>
> It is difficult to avoid the impression that a miracle confronts us here. . .
>
> The miracle of the appropriateness of the language of mathematics for the formulation of the laws of physics is a wonderful gift which we neither understand nor deserve. . .[15]

One may take issue with the phrase that we have italicized. How does one distinguish between direct and indirect suggestion? Consider, for example, Euclid's definition of a point:[16] "A *point* is that which has no part." Is this definition suggested directly by the actual world, or is it a profound abstraction from the latter? How, for that matter, does one distinguish between the

---

[13] The footnote in the original reads as follows: "This statement is quoted here from W. Dubislav's *Die Philosophie der Mathematik in der Gegenwart* [28], p. 1."

[14] It is interesting to compare Wigner's views with those of Dirac, who began a 1939 address entitled *The Relation between Mathematics and Physics* [23] as follows: "The physicist, in his study of natural phenomena, has two methods of making progress: (1) the method of experiment and observation, and (2) the method of mathematical reasoning. . . There is no logical reason why the second method should be possible at all, but one has found in practice that it does work and meets with reasonable success. This must be ascribed to some *mathematical quality in Nature*. . ."

[15] For a penetrating philosophical analysis of the problem, see the article by Steiner in [97] and his book [103].

[16] We have taken the quotation from Martin [72], p. 122, who gives references to his own sources.

actual world, and abstractions from it? Rather, how does one tell a statement about the actual world – which may be called a scientific proposition – from one about abstractions from it?

One answer to this question has been provided by Popper [84]. According to him, it is the possibility of *empirical refutation*[17] that demarcates between scientific and metaphysical propositions. A proposition cannot (or should not) be called scientific if it is not susceptible to empirical refutation.[18] Now the totality of quantitative empirical data available to mankind at any given time consists of a *finite* collection of *finite* sets of rational numbers. This fact strongly limits the inferences that can be made from such data; it excludes all conclusions that are drawn in mathematics by the use of limiting processes. Popper failed to analyze the impact of this basic limitation upon his criterion – and he has *not* been criticized by philosophers on this account.

Popper's criterion can be rendered into everyday speech as follows: A proposition can be called a *discovery* if and only if it admits the possibility of empirical refutation; otherwise – like theorems of mathematics – it can at best be called an *invention*.

Consider now the question: Is the differential calculus a discovery, or an invention [96]? A derivative involves the notion of a limit, which in turn involves an infinite data-set, which is beyond the reach of experiment or observation. This would suggest that the differential calculus is not a discovery in the sense of Popper. But then what about Newton's second law of motion, which uses the second derivative? Would any physicist be comfortable with the notion that it is an invention?

Is the calculus a discovery, or an invention? It is unlikely that any answer, or analysis, would satisfy everyone. The physicist, however, may be willing to make do with the answer that an ordered space about which one can make empirically refutable statements is densely embedded in one which is locally a differentiable manifold; the last is a statement that is not empirically refutable.

For the working physicist, the notions of *point, topology* and *completeness* come in a package. They may be logically distinct, but this logical separation is thoroughly blurred by the limitations of quantitative empirical data. One could assert that Cantor's great contribution to physics was the notion of completeness. Indeed, for the experimental physicist Cantor's nested interval theorem for complete metric spaces *is* the operational definition of a geometrical point. A geometrical point cannot be "pinpointed" in the laboratory, but, at least in classical physics, there is no limit to the "goodness" with which it can be approximated.

---

[17] The term *empirical* includes the experimental and the observational; astronomy has been called an observational, but not an experimental science. One can observe the paths of the planets, but one cannot perform experiments upon them.

[18] The authors are aware that Popper is more popular with scientists than with philosophers. In the words of a philosopher of science, "He [Popper] got everything wrong" – rather a strong statement to make about the author of *The Open Society and its Enemies* [85]. The authors are scientists, who appreciate *The Open Society*.

A

Uniformities and Uniform Completion

A *uniform space* is a set X together with a structure called a *uniformity* defined on it. This structure can be defined in three different but equivalent ways, each with its own particular advantages.[1] The first two of the following originate in the seminal work of Weil [109]; the third was defined by Tukey [108].[2]

1. In terms of a family of pseudometrics on X. This definition is used by Gillman and Jerison in their monograph on *Rings of Continuous Functions* [43], "as they [pseudometrics] provide us with a large supply of continuous functions".
2. As a filter on $X \times X$ that satisfies certain conditions, one of them being that each member of the filter contain the *diagonal* $\Delta = \{(x,x)|x \in X\}$. This definition was given in its present form by Bourbaki [12].
3. In terms of certain covers of X called *uniform covers*, in which X is covered by "sets of the same size". This definition was given by Tukey [108], and is used extensively by Isbell in his monograph [53] on what, according to its author, "might be labelled fairly accurately [as the] *intrinsic geometry of uniform spaces*".

All three definitions are given below. It should be pointed out that there are slight differences in terminology in the literature, the most important

---

[1] Willard [130] calls the second and third of these *diagonal* and *covering* uniformities, respectively, and we shall follow his terminology. In the same spirit, we shall call the first *pseudometric* uniformities. It should be borne in mind that the different names refer, not to different mathematical objects, but to different definitions of the same object.

[2] Uniformities are treated in many standard textbooks on topology. All three definitions may be found, either in the text or in the exercises, in the books by Kelley [57] and Willard [130]. The specialized monograph by Gillman and Jerison uses only pseudometric uniformities. The elementary textbook by James [55] discusses only diagonal uniformities.

H.-J. Borchers and R.N. Sen: *Mathematical Implications of Einstein–Weyl Causality*,
Lect. Notes Phys. **709**, 157–167 (2006)
DOI 10.1007/3-540-37681-X_A   © Springer-Verlag Berlin Heidelberg 2006

being that a few authors (cf. [53]) assume the Hausdorff property to be an integral part of the definition of a uniformity.

A.1 Equivalent Definitions of Uniformities

A.1.1 Pseudometric Uniformities

We begin by recalling the notion of a pseudometric.

Definition A.1.1

A pseudometric on a set X is a function $d : X \times X \to \mathbb{R}$ that satisfies the following conditions for all $x, y, z \in X$:

1. $d(x, y) \geq 0$;
2. $d(x, x) = 0$;
3. $d(x, y) = d(y, x)$; *and*
4. $d(x, z) \leq d(x, y) + d(y, z)$.

A pseudometric differs from a metric only in that $d(x, y) = 0$ need not imply $x = y$.

Example A.1.2 Let $f : X \to \mathbb{R}$ be a real-valued function on X. The function

$$d(x, y) = |f(x) - f(y)|$$

is a pseudometric on X. Note that f does not have to be continuous.

Notations A.1.3 If A is a nonempty subset of X, the *d-diameter* of A is defined to be

$$d\{A\} = \sup_{x,y \in A} d(x, y) .$$

Definition A.1.4 *A pseudometric uniformity on X is defined to be a nonempty family $\mathcal{G}$ of pseudometrics on X satisfying the following conditions:*

a) *If $d_1, d_2 \in \mathcal{G}$ then $d_1 \vee d_2 \in \mathcal{G}$, where $d_1 \vee d_2 = \sup(d_1, d_2)$.*

b) *If d is a pseudometric and $\delta > 0$ (δ is otherwise arbitrary), and there exists a $d' \in \mathcal{G}$ and $\delta' > 0$ such that $d'(x, y) < \delta' \Rightarrow d(x, y) < \delta$, then $d \in \mathcal{G}$.*

Condition b) may be expressed as follows: $d'\{A\} < \delta' \Rightarrow d\{A\} < \delta$.

A pseudometric uniformity $\mathcal{G}$ is called *Hausdorff* if for $x \neq y$ there exists a $d \in \mathcal{G}$ for which $d(x, y) \neq 0$.

A.1.2 Diagonal Uniformities

In the following, the notations E^{-1} and $F \circ F$ are to be understood as relations on X. Recall that if $E = \{(x, y)\}$ is a relation on X, i.e., a subset of $X \times X$, then the *inverse relation* E^{-1} is defined to be the subset $\{(y, x)\}$ of $X \times X$. If U and V are relations on X, then their *composition* $U \circ V$ is defined to be the set of all pairs (x, z) such that, for some y, $(x, y) \in V$ and $(y, z) \in U$.

Definition A.1.5 *The* diagonal *of a set* $X \times X$ *is the subset* $\Delta = \{(x, x) | x \in X\}$. *A* diagonal uniformity *on a set* X *is a filter* $\mathcal{E}$ *on* $X \times X$ *consisting of subsets of* $X \times X$ *called* entourages *or* surroundings *such that:*

a) *If* $E \in \mathcal{E}$ *then* $\Delta \subset E$.

b) *If* $E \in \mathcal{E}$ *then there exists an entourage* $F \in \mathcal{E}$ *such that* $F \subset E^{-1}$.

c) *If* $E \in \mathcal{E}$ *then there exists an entourage* $F \in \mathcal{E}$ *such that* $F \circ F \subset E$.

Note that $E \in \mathcal{E}$ implies that $E^{-1} \in \mathcal{E}$.

A diagonal uniformity $\mathcal{E}$ on X is called *Hausdorff* (or *separated*, or *separating*) iff

$$\bigcap_{E \in \mathcal{E}} E = \Delta \, .$$

Condition 3) of Definition A.1.5 has been described by Kelley [57] as "a vestigial form of the triangle inequality". It imposes more structure than a topology on X. Given any cover of X, one can always define a topology by taking finite intersections and arbitrary unions. Take now a family of subsets of $X \times X$ each containing the diagonal. Add all finite intersections to the family, and define a filter by taking supersets. This filter may fail to satisfy condition 3) of Definition A.1.5. See [55] for an example.

Examples A.1.6   1. On any set X, the family consisting of all supersets of the diagonal $\Delta = X \times X$ defines a uniformity called the *discrete* uniformity.
  2. On any set X, the family consisting of the single set $X \times X$ defines a uniformity called the *trivial* uniformity.

A.1.3 Covering Uniformities

We begin with a few preliminary definitions:

Definition A.1.7 *Let* $\mathcal{U}$ *and* $\mathcal{U}'$ *be covers of* X. $\mathcal{U}'$ *is said to* refine $\mathcal{U}$ *if every* U' *in* $\mathcal{U}'$ *is contained in some* U *in* $\mathcal{U}$, $U' \subset U$ *for some* $U \in \mathcal{U}$.

Definition A.1.8 *If $\mathcal{U}$ is a cover of X and $A \subset X$, then the star of A with respect to $\mathcal{U}$, written* St $(A, \mathcal{U})$, *is the union of all members of $\mathcal{U}$ that intersect A:*

$$\text{St}\,(A, \mathcal{U}) = \bigcup \{U \in \mathcal{U} \,|\, A \cap U \neq \emptyset\}\,.$$

Definition A.1.9 *Let $\mathcal{U}$ and $\mathcal{V}$ be covers of X. One says that $\mathcal{U}$ star-refines $\mathcal{V}$ (is a star-refinement of $\mathcal{V}$), iff, for each $U \in \mathcal{U}$, there is some $V \in \mathcal{V}$ such that* St $(U, \mathcal{U}) \subset V$.

Definition A.1.10 *A* covering uniformity *on a set X is a family μ of covers of X such that*

a) *If $\mathcal{V}, \mathcal{W} \in \mu$ then there exists $\mathcal{U} \in \mu$ that refines both $\mathcal{V}$ and $\mathcal{W}$.*

b) *If $\mathcal{U}$ refines $\mathcal{V}$ and $\mathcal{U} \in \mu$, then $\mathcal{V} \in \mu$.*

c) *Every element of μ has a star-refinement in μ.*

Definition A.1.11 *A* base *for a covering uniformity μ on X is any subcollection μ' of μ such that*

$$\mu = \{\mathcal{U} | \mathcal{U} \text{ covers } X \text{ and } \mathcal{U}' \text{ refines } \mathcal{U} \text{ for some } \mathcal{U}' \in \mu'\}\,.$$

A covering uniformity μ on X is called Hausdorff if for any two distinct points $x, y \in X$, there is a cover $\mathcal{U} \in \mu$ such that no element of $\mathcal{U}$ contains both x and y.

A.2 Equivalence Theorems

The results given below establish that covering uniformities are equivalent to i) pseudometric uniformities, and ii) to diagonal uniformities. The equivalence of pseudometric and diagonal uniformities follows from these.

Lemma A.2.1 *Let μ be a covering uniformity for X. Then there exists a pseudometric ρ on X such that $\mathcal{U}_\epsilon = \{U_\rho(x, \epsilon) \,|\, x \in X\}$ is a uniform cover (i.e., $\mathcal{U}_\epsilon \in \mu$) for each $\epsilon > 0$. Here $U_\rho(x, \epsilon) = \{y \,|\, y \in X, \rho(x, y) > \epsilon\}$.*

The family of pseudometrics $\{\rho_\alpha \,|\, \alpha \in A\}$ that generates a covering uniformity μ is called the *gage* $\mathcal{G}$ of the uniformity μ. The gage $\mathcal{G}$ of a uniformity μ has the properties a) and b) of Definition A.1.4. Conversely:

Theorem A.2.2 *Any collection $\mathcal{G}$ of pseudometrics on X satisfying the conditions* a) *and* b) *of Definition* A.1.4 *is a gage for some covering uniformity μ on X.*

The correspondence between gages and uniform covers is one-to-one. To state the result that establishes the correspondence between diagonal and covering uniformities, we need a definition:

Definition A.2.3 *A base $\mathcal{B}$ for a diagonal uniformity $\mathcal{E}$ on X is a filter base for $X \times X$ that satisfies conditions* a) – c) *of Definition* A.1.5.

Theorem A.2.4 *Let μ be a family of covers of X that satisfies the conditions* a)–c) *of Definition* A.1.10. *For $\mathcal{U} \in \mu$, define*

$$D_{\mathcal{U}} = \bigcup \{U \times U \mid U \in \mathcal{U}\}$$

and

$$\mathcal{B} = \{D_{\mathcal{U}} \mid \mathcal{U} \in \mu\}\,.$$

The collection $\mathcal{B}$ is a base for a diagonal uniformity $\mathcal{E}$ on X. The uniform covers of X are precisely the elements of μ.

A.3 The Uniform Topology

The three distinct but equivalent definitions of uniformities lead to three distinct but equivalent ways of defining new concepts, and stating and proving new results. We shall not give proofs, and shall confine ourself to only one definition each of the concepts that will be introduced below. Unless otherwise mentioned, our definitions will be in terms of diagonal uniformities, which will be denoted by $\mathcal{E}$. Proofs may be found in the texts by Kelley [57], Willard [130] or James [55].

We shall sometimes view subsets of $X \times X$ as relations. Accordingly, if $E \subset X \times X$, we shall write:

Notations A.3.1

$$E[x] = \{y \mid (x, y) \in E\}.$$
$$E[A] = \bigcup_x E[x], \quad x \in A \subset X.$$

Definition A.3.2 *If $(X, \mathcal{E})$ is a uniform space, the topology $\mathcal{T}$ of the uniformity $\mathcal{E}$, or the* **uniform topology,** *is the topology in which the family of sets $\mathcal{T}_x = \{E[x] \mid E \in \mathcal{E}\}$ is a neighbourhood base at $x \in X$.*

The topology of the discrete uniformity is discrete; that of the trivial uniformity is indiscrete.

Proposition A.3.3 *The topology $\mathcal{T}$ of the uniformity $\mathcal{E}$ is Hausdorff if and only if $\mathcal{E}$ is Hausdorff.*

On a uniformizable space (the definition of uniformizability is given below, in Definition A.5.1), there may be different uniformities that gives rise to the same topology. However, this cannot happen in a compact Hausdorff space:

Theorem A.3.4 *If X is a compact Hausdorff space, then the neighbourhood filter of the diagonal ΔX in $X \times X$ defines a uniformity on X. The topology of this uniformity coincides with the original topology, and no other uniformity on X has this property.*

Definition A.3.5 (Relative uniformity) *If $(X, \mathcal{E}_X)$ is a uniform space and Y a subset of X, then the uniformity defined on Y by the family*

$$\mathcal{E}_Y = \{E \cap Y \times Y \,|\, E \in \mathcal{E}_X\}$$

is called the relativization *of $\mathcal{E}_X$ to Y, or the* relative uniformity *for Y. The pair $(Y, \mathcal{E}_Y)$ is called a* uniform subspace *of $(X, \mathcal{E}_X)$.*

Use of the term *relativization* in the theory of uniformities is consistent with its usage in general topology.

A.4 Uniform Continuity and Equivalence

Definition A.4.1 *Let $\mathcal{E}$ and $\mathcal{K}$ be diagonal uniformities on X and Y respectively. A function $f : X \to Y$ is said to be* uniformly continuous *iff for each $K \in \mathcal{K}$, there is some $E \in \mathcal{E}$ such that $(x_1, x_2) \in E \Rightarrow f(x_1, x_2) \in K$.*

It follows easily that every uniformly continuous function is continuous.
The notion of uniform equivalence is akin to that of topological equivalence or homeomorphism, but there appears to be no shorter term for it.

Definition A.4.2 *The uniform spaces X and Y are said to be* uniformly equivalent *if there exists a bijection $\phi : X \to Y$ such that both ϕ and ϕ^{-1} are uniformly continuous.*

The notion of a Cauchy sequence extends to uniform spaces, and a uniformly continuous map sends Cauchy sequences to Cauchy sequences, which is a result that we shall use several times.

Definition A.4.3 *The sequence of points $\{x_n\}$ in a uniform space X is called a* Cauchy sequence *if for each entourage E of X there is an integer N such that $(x_m, x_n) \in E$ whenever $m, n \geq N$.*

Proposition A.4.4 *Let X, Y be uniform spaces and $f : X \to Y$ a uniformly continuous map. If $\{x_n\}_{n \in \mathbb{N}}$ is a Cauchy sequence in X, then $\{f(x_n)\}_{n \in \mathbb{N}}$ is a Cauchy sequence in Y.*

A.5 Uniformizability of a Topological Space

We now come to the notion of uniformizability, which is a property of topological spaces.

Definition A.5.1 *A topological space X is said to be* uniformizable *if there is a uniformity $\mathcal{E}$ on X such that the topology of this uniformity coincides with the topology of X.*

Theorem A.5.2 *A topological space is uniformizable if and only if it is completely regular.*[3]

A.5.1 Metrizability of a Uniform Space

A metric on X defines a uniformity on X via a cover consisting of open balls of radius $\leq \epsilon$. This uniformity is called the *metric uniformity* on X.

Definition A.5.3 *A uniformity $\mathcal{E}$ on X is said to be* metrizable *if there exists a metric d on X such that the uniformity induced by d coincides with $\mathcal{E}$.*

Theorem A.5.4 *A uniformity is metrizable if and only if it is Hausdorff and has a countable base.*

An ordered space is always Hausdorff, but *does not have to be second countable.*

Remark A.5.5 If a uniformity is metrizable, so is the uniform topology it generates. In the opposite direction, there is a surprise: metrizability of the uniform topology does *not* imply that the uniformity itself is metrizable. See [130] for an example.

A.6 Uniform Completion

We begin by recalling the definition of a Cauchy filter in a uniform space.

Definition A.6.1 (Cauchy filter) *A filter $\mathcal{F}$ in the uniform space $\{X, \mathcal{E}\}$ is called a* Cauchy *filter if for each entourage E of the uniformity there exists a member W of $\mathcal{F}$ such that $W \times W \subset E$.*

Remark A.6.2 It is sufficient for the Cauchy condition to be satisfied for all members of a base for the uniformity.

Next, we recall the following result about sequences in a uniform space:

---

[3] Recall that we use the term *completely regular* to denote a space in which a closed set and a point disjoint from it can be separated by an Urysohn function; we do not require one-point sets to be closed. We call a completely regular space in which one-point sets are closed a *Tychonoff* space.

Proposition A.6.3 *Let $\{x_n\}$ be a sequence of points in the uniform space X. If $\{x_n\}$ converges in the uniform topology then $\{x_n\}$ is a Cauchy sequence.*

The analogue of the above for filters is:

Proposition A.6.4 *Let $\mathcal{F}$ be a filter on the uniform space X. If $\mathcal{F}$ converges in the uniform topology then $\mathcal{F}$ is a Cauchy filter.*

The key definition is the following:

Definition A.6.5 *A uniform space is called* **complete** *if and only if every Cauchy filter in the space converges to a point in the space.*

Also:

Definition A.6.6 *A uniform space M is called* sequentially complete *if every Cauchy sequence of points in M converges.*

Theorem A.6.7 *A complete uniform space is sequentially complete.*

A uniform space that is not complete may be completed (by identifying Cauchy filters with points). Details may be found in [57], [130] or [55].

Definition A.6.8 (Uniform completion of a uniform space) *A uniform completion of a uniform space $(X, \mathcal{E})$ is a pair $(f, (X^\star, \mathcal{E}^\star))$ where $(X^\star, \mathcal{E}^\star)$ is a complete uniform space and f is a uniform embedding of X as a dense subspace of $X^\star$.*

   Propositions A.6.3 and A.6.4 provide the essentials of the connection between the completion of a uniformity and the completion of a metric space. Note that there exist uniformities that do not arise from any metric. For details, see [55].

A.6.1 Complete Uniformizability: Shirota's Theorem

Finally, we come to another property of topological spaces, called complete uniformizability:

Definition A.6.9 *A topological space X is called* completely uniformizable *if there exists a uniformity in which X is complete, and which induces the topology of X.*

Theorem A.6.10 (Shirota's theorem) *A topological space X is completely uniformizable iff:*

a) *X is Tychonoff.*

b) *Every closed discrete subspace of X has nonmeasurable cardinal, and:*

c) *X is realcompact.*[4]

---

[4]   We shall not give either the proper definition of realcompactness, or the motivation for it, but shall content ourselves with quoting the following result: *Theorem:* A topological space is realcompact iff it can be embedded as a closed subspace in a product of real lines. The interested reader may consult [43].

The cardinal m of the power of a set S is called *nonmeasurable* if every countably additive two-valued measure (taking the values 0 and 1) defined on all subsets of S and vanishing on singletons vanishes.[5] Observe that the cardinal $\aleph_0$ is nonmeasurable.

Neither the existence nor the nonexistence of measurable cardinal numbers is provable in ZFC.[6] Moreover, assuming that ZFC is consistent, it is not possible to prove that "ZFC + there exists a measurable cardinal" is consistent. It follows that in mathematics based on ZFC the cardinality condition in Shirota's theorem may be dropped. We refer the reader to [39] for an introduction to "beyond ZFC" and to [27] for more details, without the use of forcing.

Shirota's theorem has been called by Isbell [53] "the first deep theorem in uniform spaces".

A.6.2 Total Boundedness

There is a useful connection between completeness and compactness in the class of uniform spaces called *totally bounded* or *precompact*:

Definition A.6.11 *A covering uniformity μ on X is said to be* totally bounded (*or* precompact) *if μ has a base consisting of finite covers. Equivalently, a diagonal uniformity $\mathcal{E}$ on X is totally bounded iff, for each $E \in \mathcal{E}$, there is a finite cover $\{U_1, \ldots U_n\}$ of X such that $U_k \times U_k \subset E$ for each k.*

If X is equipped with a totally bounded uniformity, it is called a *totally bounded* (or *precompact*) uniform space.

Theorem A.6.12 *A uniform space is compact iff it is complete and totally bounded.*

That is, a precompact uniform space is compact iff it is complete.

A.7 Properties of Hausdorff Uniformities

The order uniformity is Hausdorff, and Hausdorff uniformities have many pleasant properties. We list below those that we shall exploit.

Theorem A.7.1 *Every uniform space is uniformly isomorphic to a dense subspace of a complete uniform space. Each Hausdorff uniform space is uniformly isomorphic to a dense subspace of a complete Hausdorff uniform space.*

---

[5] The question whether such a measure, which does not vanish identically, exists is known as *Ulam's problem*.

[6] The standard version of mathematics is based on the Zermelo-Fraenkel axioms for set theory (denoted ZF), plus the axiom of choice (denoted C), and is sometimes called ZFC. See [48] for a brief but lucid account, or [39] for a fuller account.

For proofs, see [55], [57] or [130].

The completion $(X^\star, \mathcal{E}^\star)$ is Hausdorff if $(X, \mathcal{E})$ is a Hausdorff uniform space. For this case, Theorem A.7.1 can be stated as follows:

Theorem A.7.2 *The uniform completion of a (Hausdorff) uniform space is itself a (Hausdorff) uniform space.*

Furthermore, the uniform completion of a Hausdorff uniform space is essentially unique:

Theorem A.7.3 *Let X_1, X_2 be complete Hausdorff uniform space, and let A_1, A_2 be dense subsets of X_1, X_2 respectively. If A_1 and A_2 are uniformly equivalent then so are X_1 and X_2.*

Theorem A.7.4 *A compact Hausdorff space admits of a unique uniformization.*

We conclude this section with the following result (which we have exploited repeatedly in Chap. 6).

Theorem A.7.5

   1) *A closed subset of a complete uniform space is complete.*

   2) *A complete subspace of a Hausdorff uniform space is closed.*

A.8 Inequivalent Uniformities

A Tychonoff space is uniformizable. A uniformization of a Hausdorff topological space is Hausdorff, so that a uniformization of a Tychonoff space is Hausdorff. Therefore any uniformization of a Tychonoff space has an essentially unique (Hausdorff) uniform completion. However, a Tychonoff space can admit of inequivalent uniformities which have very different completions, as the following example (rather, class of examples) illustrates.

Let X be a Tychonoff space. Let $C(X)$ denote the space of continuous real-valued functions on X, and $C^*(X)$ the space of *bounded* real-valued functions on X. Any $f \in C(X)$ or defines a pseudometric on X. The family of these pseudometrics defines a (pseudometric) uniformity on X, denoted by $\mathcal{C}(X)$. Similarly, the functions $f \in C^*(X)$ define a uniformity on X, which is denoted by $\mathcal{C}^*(X)$. (For details, see [43].) If X is not compact, these two uniformities are inequivalent.

The Stone-Čech compactification βX of X is carried out using $C^*(X)$. (See, for example, [130].) If one carries out exactly the same construction, but using $C(X)$ instead of $C^*(X)$, one obtains a space which is denoted by υX and is called the *Hewitt realcompactification* or the *Nachbin completion* of X. The following theorem holds:

Theorem A.8.1 *Let X be a completely regular space.*

1. *The completion of X in the uniformity $\mathcal{C}(X)$ is $(\upsilon X, \mathcal{C}(\upsilon X))$.*

2. *The completion of X in the uniformity $\mathcal{C}^*(X)$ is $(\beta X, \mathcal{C}(\beta X))$.*

For a proof, see [43]. The reader will also find there (in the exercises) conditions under which a Tychonoff space has a unique uniformization.

The spaces υX and βX are related as follows:

Proposition A.8.2
$$X \subset \upsilon X \subset \beta X \ .$$

We have not explained either the origins or the full meaning of the term *realcompactification*, as to do so would have taken us too far afield. The interested reader will find a detailed treatment of the subject in the monograph [43] by Gillman and Jerison.

This concludes our summary of uniformities and uniform completion.

B

Fibre Bundles and *G*-Structures

In this appendix, we shall recapitulate a few key definitions and results from the theory of fibre bundles and *G*-structures. We shall define *G*-structures using the concept of coordinate transformations on the base space with values in a group, which was made an integral part of his definition of fibre bundles by Steenrod [102]. Fibre bundles enjoy a crucial property known as the homotopy lifting property (the covering homotopy theorems; [102], Sect. 11), and in topological applications it is often this property that is essential. Accordingly, in the topological literature one often comes across definitions of fibre bundles that do not refer to coordinate transformations on the base space with values in a group, but in geometry one has to ask for more.

The material of Sect. B.1 provides the background required for Sect. 8.5. The material of Sect. B.2 is meant to supplement that of Sect. 2.2.

B.1 Fibre Bundles

The definition that follows is taken from Steenrod (*op. cit.*). Variants that are suitable for geometry may be found in [62] or [104].

B.1.1 Coordinate Transformations on the Base Space

When global coordinate systems do not exist, there are no global coordinate transformations. The following gives a local version.

Let X be a topological space and $\{V_j\}$, $j \in J$ an open cover of X (J is an indexing set), and G a topological group. A family of *continuous* maps $\{g_{ji}\}$

$$g_{ji} : V_i \cap V_j \to G , \tag{B.1}$$

is called a *system of coordinate transformations on X with values in G* if it satisfies the following conditions:

H.-J. Borchers and R.N. Sen: *Mathematical Implications of Einstein–Weyl Causality*,
Lect. Notes Phys. **709**, 169–173 (2006)
DOI 10.1007/3-540-37681-X_B   © Springer-Verlag Berlin Heidelberg 2006

$$g_{ij}(x)g_{jk}(x) = g_{ik}(x) \quad \forall\, x \in V_i \cap V_j \cap V_k \ ,$$
$$[g_{ik}(x)]^{-1} = g_{ki}(x) \quad \forall\, x \in V_i \cap V_k \ . \tag{B.2}$$

It follows from the above that

$$g_{ii}(x) = e \ \forall\, i \in J, \ x \in V_i, \tag{B.3}$$

where e is the identity of G.

B.1.2 Fibre Bundles

The topological notion of a fibre bundle is a generalization of the notion of the topological product $X \times Y$ of the spaces X and Y. A *fibre bundle* $\mathcal{B} = \{B, X, Y, p, G\}$ is a collection as follows: X, Y and B are topological spaces called respectively the *base space*, *fibre* and *total space*. G is a topological group called *the group of the bundle* which acts effectively[1] on Y. Finally, $p : B \to X$ is a surjective map called the *projection*. The space B is locally, but not necessarily globally, a product (one sometimes says that B is *locally trivial*). One may picture B as being glued together from "local products" $V_j \times Y$, where $V_j \subset X$ is an open set. $V_j \times Y$ may, in turn be viewed as a collection of "fibres" $\{x\} \times Y$, where $x \in V_j$, whence the term fibre bundles. The glueing-together is done as follows. One demands that there exist an open cover $\{V_j\}$, $j \in J$ of X such that, for each $j \in J$, there is a homeomorphism

$$\phi_j : V_j \times Y \to p^{-1}(V_j) \tag{B.4}$$

which *preserves fibres*, i.e., satisfies

$$p\phi_j(x, y) = x \ \forall\, x \in V_j, \ y \in Y \ .$$

Then for each $j \in J$ and each $x \in V_j$, the expression

$$\phi_{j,x}(y) = \phi_j(x, y)$$

maps Y onto the fibre $p^{-1}(x)$ over x,

$$\phi_{j,x} : Y \to p^{-1}(x),$$

and the composite map

$$\phi_{j,x}^{-1} \circ \phi_{i,x} : Y \to Y$$

is a homeomorphism of Y. This homeomorphism is required to coincide with the operation of an element g of G on Y. Define now

$$g_{ji}(x) = \phi_{j,x}^{-1} \circ \phi_{i,x}(x) \ . \tag{B.5}$$

---

[1] That is, $g \cdot y = y$ for all $y \in Y$ implies that $g = e$ (the group is assumed to act from the left).

The g_{ji} so defined are required to form a system of coordinate transformations on X with values in G.[2]

In Steenrod's definition, X, Y, G and the $\{V_j, g_{ji}\}$ are taken as given; the total space B and the projection p are constructed. The construction, for the case of interest, is carried out explicitly in Sect. 8.5 and will not be repeated here. The result is a fibre bundle with a coordinate system; the latter is eliminated in the standard fashion. For details, see [102].

Let $\mathcal{B} = \{B, X, Y, p, G\}$ and $\mathcal{B}' = \{B', X', Y, p', G\}$ be two bundles with the same fibre and group. A map $h : B \to B'$ is called a *bundle map* if it satisfies the following conditions:

1. It preserves fibres, i.e., maps the fibre Y_x over $x \in X$ homeomorphically onto a fibre $Y_{x'}$, thus inducing a continuous map $\bar{h} : X \to X'$ such that

$$p' \circ h = \bar{h} \circ p .$$

2. Let $x' = \bar{h}(x)$. If $x \in V_j \cap \bar{h}^{-1}(V'_k)$ and $h_x : Y_x \to Y_{x'}$ is the map induced by h, then the map

$$\bar{g}_{kj}(x) = \phi'^{-1}_{k,x'} \circ h_x \circ \phi_{j,x}$$

of Y into Y coincides with the operation of an element of G.

3. The map

$$\bar{g}_{kj} : V_j \cap \bar{h}^{-1}(V'_j) \to G$$

defined above is continuous.

These maps are sometimes called fibre-preserving maps in the literature.

B.1.3 Reduction of the Group of the Bundle

If the cover $\{V_j\}$ of X contains a subcover $\{V_\alpha\}_{\alpha \in A}$ such that the $g_{\alpha\beta}$ take their values in a closed subgroup H of G whenever $\alpha, \beta \in A$, restriction to this subcover is said to effect a *reduction of the group of the bundle* (from G to H).

All of the above considerations carry over to smooth fibre bundles if topological spaces are replaced by smooth manifolds, continuous maps by smooth maps, topological groups by Lie groups and continuous actions by smooth actions. Fibre bundles in differential geometry are assumed to be smooth.

B.1.4 Tangent Bundles

Let M be a smooth manifold of dimension n and $T_x(M)$ be the tangent space to it at $x \in M$. Then $T_x(M)$ is isomorphic to $\mathbb{R}^n$. It can be shown that M, together with the tangent spaces at all its points, is a fibre bundle with fibre $Y = \mathbb{R}^n$ and group $GL(n, \mathbb{R})$. This bundle is called the *tangent bundle* and denoted by $T(M)$. Details may be found in Steenrod's book [102], or in [62], [75] or [104].

---

[2] In relativity theory, the symbol g_{ij} usually denotes components of the pseudo-Riemannian metric. This is clearly not the case over here.

B.2 G-structures on Differentiable Manifolds

Many of the important local geometrical structures that can be defined on a differentiable manifold[3] fall into two classes. They are *G-structures*, defined by Chern [19], and *connections*, first defined under the name of parallel transport by Levi-Civita [70] and later subsumed under the general notion elaborated by Ehresmann (see [62] or [104]). We shall recapitulate the notion of G-structures in the following; we shall not have occasion to use the notion of connections. We remind the reader that we shall only be considering smooth structures.

It is a fundamental result in the theory of fibre bundles (see, for example, [102]) that *the group of a bundle can always be reduced to a maximal compact subgroup*.

We now make the following definition:

Definition B.2.1 (G-structures:)

A G-structure on a manifold M with group G is a reduction of the group $GL(n, \mathbb{R})$ of the tangent bundle $T(M)$ to the group G.

The following examples are familiar under other descriptions:

Examples B.2.2

1. **Riemannian metric:** A Riemannian metric on M is a smooth assignment of positive definite quadratic forms on tangent spaces $T_x(M)$ at the points x of M. This is equivalent to reducing the group of the tangent bundle to the orthogonal group $O(n)$, for the latter is the group of linear transformations that leave positive definite quadratic forms on $T_x(M)$ invariant. Since $O(n)$ is the maximal compact subgroup of $GL(N, \mathbb{R})$, this reduction is always possible; a differentiable manifold always admits a Riemannian metric. A Riemannian metric on M is an $O(n)$-structure on M.

2. **Pseudo-Riemannian metric:** A pseudo-Riemannian metric on the n-dimensional manifold M is a smooth assignment of indefinite quadratic forms of signature $n-2$ at the tangent spaces $T_x(M)$ at points x of M. This is equivalent to reducing the group of the tangent bundle to the Lorentz group $O(1, n-1)$, for the latter is the group of linear transformations that leave the Minkowski form $x_0^2 - x_1^2 - \ldots - x_{n-1}^2$ invariant. A pseudo-Riemannian metric on M is an $O(1, n-1)$-structure on M; it is also known as a *Lorentz structure*. Among compact two-dimensional manifolds, only two admit a Lorentz structure: the torus and the Klein bottle. By contrast, every compact three-manifold admits a Lorentz structure [102].

3. **Conformal structure:** Denote by (u, v) the (indefinite) Minkowski form on $\mathbb{R}^n$. A Lorentz transformation Λ leaves this form invariant: $(\Lambda u, \Lambda v) = (u, v)$. One may consider, instead of the Λ's, transformations $T \in GL(n, \mathbb{R})$

---

[3]   We assume that a manifold is Hausdorff, second countable and paracompact, by definition. See footnote 8 on p. 150.

that multiply the Minkowski form by a nonzero scalar: $(Tu, Tv) = \lambda(T)(u, v)$ for all $u, v \in \mathbb{R}^n$, where $\lambda(T)$ depends only on T. The *conformal group* is the group G of these transformations. A *conformal structure* on an n-dimensional pseudo-Riemannian manifold M is a reduction of the group of $T(M)$ to G.

Remark B.2.3 The definition of conformal structure given above is the one generally found in the physics literature; see, for example, [46], pp. 15–19 or [54], pp. 642–645. Conformal transformations thus defined preserve both "spherical" and "hyperbolic" angles. In the mathematical literature, one usually considers only Riemannian metrics and spherical angles. See, for example, [104]. The definition used in physics includes the one used in mathematics as a special case. The concept of conformal transformations used earlier in Sect. 2.2.3 was that of the physicist.

C

The Axioms and Special Assumptions

For ease of reference, the axioms, the nontriviality assumptions (3.1.10), the definition of a D-set and the special assumptions of Chaps. 5 and 6 are collected together in the following. The verbal explanations that follow the formal definitions a)–f) of (4.2.1) have been omitted. The entries below are numbered as they are in the text. Recall that βC is the subset of the cone C which, in a D-set, is seen to coincide with the boundary of C after the topology is introduced (Sects. 3.2 and 3.2.1).

Axiom 3.1.2 (The order axiom, p. 17)

a) If $x, y \in l$ and $x \neq y$, then either $x <^l y$ or $y <^l x$; if $x <^l y$ and $y <^l$, then $x = y$.

b) If $x, z \in l$, $x \neq z$, $x <^l z$, then $\exists\, y \in l$ such that $x <^l y <^l z$, $x \neq y$, $y \neq z$.

c) If $y \in l$, then $\exists\, x, z \in l$ such that $x <^l y <^l z$, $x \neq y$, $y \neq z$.

d) If $x, y \in l^1 \cap l^2$, then $x <^{l^1} y \Leftrightarrow x <^{l^2} y$.

$\square$

Assumptions 3.1.10 (Nontriviality assumptions, p. 21)

a) M is nonempty, and does not consist of a single point.

b) M does not consist of a single light ray.

c) M is l-connected.

H.-J. Borchers and R.N. Sen: *Mathematical Implications of Einstein–Weyl Causality*,
Lect. Notes Phys. **709**, 175–177 (2006)
DOI 10.1007/3-540-37681-X_C
© Springer-Verlag Berlin Heidelberg 2006

Axiom 3.1.12 (The identification axiom, p. 21)

If l and l' are distinct light rays and $a \in S \equiv l \cap l'$, then there exist $p, q \in l$ such that $p <^{ll} a <^{ll} q$, and $l(p,q) \cap S = \{a\}$. Similarly for l'.

$\square$

Axiom 3.2.8 (The cone axiom, p. 25)

$$C_x^+ \cap C_x^- = \{x\} \ \forall \ x \in M .$$

$\square$

Definition 4.2.1 (p. 32)

A subset U of M will be called a *D-set* (from the German *Durchschnittseigenschaft*) iff it fulfills the following conditions:

a) $x, y \in U \Rightarrow I[x,y] \subset U$.

b) For every $x \in U$ and every $l \ni x$, there exist points $p, q \in l \cap U$ such that $p <^{ll} x <^{ll} q$.

c) If $y \in U, r \in \tau C_y^- \cap U$ and $l_r \ni r$, then

$$l_r^+ \cap \{\beta C_y^- \backslash \{y\}\} \cap U \neq \emptyset ,$$

and the same for reversed order.

d) If $x \in U$ and $l \cap \beta C_x^+ \cap U$ contains two distinct points a, b, then

$$x \in l_{a,b} \cap C_x^+ \cap U \subset \beta C_x^+ \cap U ,$$

and the same for reversed order. (Remark: This part is called the *convexity axiom*.)

e) If $a, b \in U$ and $\lambda(a,b)$, then the ray $l_{a,b}$ is unique.

f) If $x \in U$, then there pass at least two distinct light rays through x.

$\square$

Axiom 4.5.1 (The local structure axiom, p. 45)

The ordered space M satisfies the following axiom: For each $x \in M$, there exists a D-set U_x such that $x \in U_x \subset M$.

$\square$

Assumption 5.3.3 (Overlapping cover assumption, p. 61)

The ordered space M will be assumed to be such that every light ray has an overlapping cover.

This assumption is not needed in order complete spaces.

Assumption 6.4.1 (Local precompactness assumption, p. 72)

From now on (unless the contrary is stated explicitly) all ordered spaces M will be assumed to be locally precompact. That is, the uniform subspace $(D, \mathcal{E}_D)$ will, by assumption, be totally bounded for every D-set $D \subset M$. Here $\mathcal{E}_D$ is the order uniformity on D.

Assumption 6.5.1 (First countability assumption, p. 76)

From now on, all ordered spaces M will be assumed to satisfy the first axiom of countability, unless the contrary is stated explicitly.

Assumption 6.6.1 (Dimensionality assumption, p. 80)

There are infinitely many light rays through any point of M.

This assumption is not needed in locally Minkowski spaces.

References

1. A. D. Alexandrov (1959). *Filosofskoe Soderzhanie i Znachenie Teorii Otnositel'nosti*, *Voprosy Filosofii*, **1**, 67–84.
2. A. Ashtekar (2004). Background Independent Quantum Gravity: A Status Report, *Classical and Quantum Gravity*, **21**, R53–R152.
3. M. F. Atiyah (1979). *The Geometry of Yang-Mills Fields*, Lezioni Fermiane, Scuola Normale Superiore, Pisa.
4. L. Auslander and R. E. MacKenzie (1963). *Introduction to Differentiable Manifolds*, McGraw-Hill, New York.
5. A. Avishai and H. Ekstein (1973). Presymmetry of Classical and Relativistic Fields, *Phys. Rev.*, **D 7**, 983–991.
6. B. Bahns, S. Doplicher, K. Fredenhagen and G. Piacitelli (2005). Field Theory on Noncommutative Spacetimes: Quasiplanar Wick Products, *Phys. Rev.*, **D 71**(2), p. 12.
7. G. Birkhoff and M. K. Bennett (1988). Felix Klein and his. "Erlanger Programm": History and Philosophy of Modern Mathematics, *Philos. Sci.*, **X**, Minneapolis, MN, pp. 145–176.
8. J. D. Bjorken and S. Drell (1967). *Relativistic Quantum Fields*, McGraw-Hill, New York.
9. L. Bombelli, J. Lee and R. D. Sorkin (1987). Space-time as a Causal Set, *Phys. Rev. Lett.*, **59**, pp. 521–524.
10. H.-J. Borchers and R. N. Sen (1990). Theory of Ordered Spaces, *Commun. Math. Phys.*, **132**, pp. 593–611.
11. H.-J. Borchers and R. N. Sen (1999). Theory of Ordered Spaces, II. The Local Differential Structure, *Commun. Math. Phys.*, **204**, pp. 475–492.
12. N. Bourbaki (1960). *Topologie Générale*, Chaps. 1 et 2, 3rd edition, Actualites Sci. et Ind. **1142**, Hermann, Paris. English translation, *General Topology*, Parts 1 and 2, Addison-Wesley, Reading, MA, 1966.
13. J. P. Bourguignon and H. B. Lawson Jr. (1982). *Yang-Mills Theory, its Physical Origins and Differential Geometric Aspects*, Ann. of Math. Studies 102, Princeton University Press, Princeton.
14. O. Bratteli and D. W. Robinson (1979). *Operator Algebras and Quantum Statistical Mechanics I*, Springer-Verlag, New York.
15. F. E. Browder, Ed. (1976). *Mathematical Developments arising from Hilbert Problems*, vol. 1 (Proceedings of Symposia in Pure Mathematics, vol. 28), American Mathematical Society, Providence, RI.

180   References

16. G. Cantor (1883). *Grundlagen einer allgemeinen Mannigfaltigkeitslehre. Ein mathematisch-philosophischer Versuch in der Lehre des Unendlichen*, Teubner, Leipzig, p. 47. This book appeared originally as a paper: Ueber unendliche, lineare Punktmannigfaltigkeiten, V, *Ann. Math.*, **21**, pp. 545–591 (1883). Reprinted in [17].

17. G. Cantor (1932). *Gesammelte Abhandlungen mathematischen und philosophischen Inhalts*, Ed. E. Zermelo, J. Springer, Berlin. Reprinted Olms, Hildesheim, 1966. Reference [16] appears on pp. 165–208 of this collection.

18. K. Chandrasekharan, Ed. (1986). *Hermann Weyl, 1885–1985*, Centenary Lectures delivered by C. N. Yang, R. Penrose and A. Borel at the ETH Zürich, Springer-Verlag, Berlin.

19. S. S. Chern (1966). The Geometry of G-Structures, *Bull. Amer. Math. Soc.*, **72**, pp. 167–219.

20. J. W. Dauben (1979). *Georg Cantor: His Mathematics and Philosophy of the Infinite*, Princeton University Press, Princeton, NJ, U.S.A. (paperback, 1990).

21. W. Deppert, K. Hübner, A. Oberschelp and V. Weidemann, Eds. (1988). *Exakte Wissenschaften und ihre philosophische Grundlegung (Exact Sciences and their Philosophical Foundations)*, Vorträge des Internationalen Hermann-Weyl-Kongresses (Kiel, 1985), Peter Lang AG, Frankfurt.

22. R. L. Devaney (1989). *An Introduction to Chaotic Dynamical Systems*, Second edition, Addison-Wesley, Reading, MA, U.S.A.

23. P. A. M. Dirac (1938–39). The relation between mathematics and physics, *Proc. Roy. Soc. (Edinburgh)*, **59**, pp. 122–129.

24. J. Dixmier (1964). *Les C^*-algèbres et leurs representations*, Gauthier-Villars, Paris.

25. S. K. Donaldson (1983). An application of gauge theory in 4-dimensional topology, *J. Diff. Geom.*, **18**, pp. 279–313.

26. S. Doplicher, K. Fredenhagen and J. Roberts (1994). Space-Time Quantization Induced by Classical Gravity, *Phys. Lett.*, **B 331**, pp. 39–44.

27. F. R. Drake (1974). *Set Theory: An Introduction to Large Cardinals*, vol. 76 of "Studies in Logic and the Foundations of Mathematics", North-Holland, Amsterdam.

28. W. Dubislav (1932). *Die Philosophie der Mathematik in der Gegenwart*, Junker und Dünnhaupt Verlag, Berlin.

29. M. Dütsch and K.-H. Rehren (2002). A Comment on the Dual Field in the Scalar AdS-CFT Correspondence, *Lett. Math. Phys.*, **62**, pp. 171–184.

30. J. Dugundji (1966). *Topology*, Allyn and Bacon, Boston, MA.

31. F. W. Dyson, A. S. Eddington and C. Davidson (1920). A Determination of the Deflection of Light by the Sun's Gravitational Field from Observations made at the Total Eclipse May 29, 1919, *Phil. Trans. Roy. Soc.*, **220 A**, pp. 291–333.

32. J. Ehlers, F. A. E. Pirani and A. Schild (1972). The Geometry of Free Fall and Light Propagation, article in [79], pp. 63–84.

33. A. Einstein (1911). Über den Einfluss der Schwerkraft auf die Ausbreitung des Lichtes, *Ann. Phys.*, **35**, pp. 898–908.

34. A. Einstein (1916). Die Grundlage der allgemeinen Relativitätstheorie, *Annalen der Physik*, **49**, pp. 769–822.

35. L. P. Eisenhart (1964). *Riemannian Geometry*, Princeton University Press, Princeton, NJ, USA, Fifth printing.

36. H. Ekstein (1967). Presymmetry, *Phys. Rev.*, **153**, pp. 1397–1402.

37. H. Ekstein (1969). Presymmetry II, *Phys. Rev.*, **184**, pp. 1315–1337.
38. A. A. Fraenkel (1953). *Abstract Set Theory*, in: "Studies in Logic and the Foundations of Mathematics", North-Holland, Amsterdam.
39. A. A. Fraenkel, Y. Bar-Hillel and A. Levy (1973). *Foundations of Set Theory*, 2nd revised edition, vol. 67 of "Studies in Logic and the Foundations of Mathematics", North-Holland, Amsterdam (reprinted 2001).
40. M. H. Freedman (1984). There is no room to spare in four-dimensional space, *Notices Amer. Math. Soc.*, **31**, pp. 3–6.
41. C. F. Gauss (1827). *Disquitiones circa superficias curvas*. See *Werke*, vol. IV, pp. 217–258, published by the Königlichen Gesellschaft der Wissenschaften von Göttingen, 1873. Translated into German by A. Wagerin, in C. F. Gauss: *Allgemeine Flächentheorie*, Verlag Wilhelm Engelmann, Leipzig, pp. 3–51, 1912.
42. R. Gompf (1983). Three exotic $\mathbb{R}^4$s and other anomalies, *J. Diff. Geom.*, **18**, pp. 317–328.
43. L. Gillman and M. Jerison (1960). *Rings of Continuous Functions*, van Nostrand Reinhold, New York, NY.
44. A. K. Guts (1982). Axiomatic Relativity Theory, *Russian Math. Surveys*, **37**(2), pp. 41–89.
45. A. K. Guts and A. V. Levichev (1984). On the Foundations of Relativity Theory, *Soviet Math. Dokl.*, **30**, pp. 253–257.
46. R. Haag (1993). *Local Quantum Theory*, Springer-Verlag, Berlin.
47. J. Hadamard (1923). *Lectures on Cauchy's Problem in Linear Partial Differential Equations*, Yale University Press, New Haven. Reprinted by Dover Publications, New York, 1952.
48. A. G. Hamilton (1978). *Logic for Mathematicians*, Cambridge University Press, Cambridge.
49. D. Hilbert (1902). *Foundations of Geometry*, translated from the German by E. J. Townsend, The Open Court Publishing Company, La Salle, IL.
50. M. W. Hirsch (1976). *Differential Topology*, Springer-Verlag, New York.
51. E. P. Hubble (1929). A Relation between Distance and Radial Velocity among Extra-Galactic Nebulae, *Proc. Nat. Acad. Sci. USA*, **19**, pp. 168–173.
52. W. Hurewicz and H. Wallman (1948). *Dimension Theory*, Princeton University Press, Princeton.
53. J. R. Isbell (1964). *Uniform Spaces*, American Mathematical Society, Providence, RI.
54. C. Itzykson and J.-B. Zuber (1980). *Quantum Field Theory*, International Series in Pure and Applied Physics, McGraw-Hill, New York. International Edition, McGraw-Hill, Singapore, 1985.
55. I. M. James (1999). *Topologies and Uniformities*, Springer-Verlag, London.
56. P. Jordan (1952). *Schwerkraft und Weltall*. Vieweg & Sohn, Braunschweig, 1955.
57. J. L. Kelley (1955). *General Topology*, van Nostrand, New York.
58. M. A. Kervaire (1960). A manifold which does not admit of any differentiable structure, *Comment. Math. Helv.*, **34**, pp. 257–270.
59. H. Kneser (1958). Analytische Struktur und Abzählbarkeit, *Ann. Acad. Sci. Fenn.*, **A I**(251/5), pp. 3–8.
60. H. Kneser (1958). Sur les variétés connexes de dimension 1, *Bull. Soc. Math. de Belgique*, **X**, pp. 19–25.
61. H. Kneser and M. Kneser (1960). Reell-analytische Strukturen der Alexandroff-Geraden, *Arch. d. Math.*, **9**, pp. 104–106.

62. S. Kobayashi and K. Nomizu (1963). *Foundations of Differential Geometry*, vol. I, Wiley-Interscience, New York.
63. W. Koch and D. Puppe (1968). Differenzierbare Strukturen auf Manninfaltigkeiten ohne abzählbare Basis, *Arch. d. Math.*, **19**, pp. 95–102.
64. A. A. Kosinski (1992). *Differential Manifolds*, Academic Press, San Diego, CA.
65. H. A. Kramers (1927). La diffusion de la lumiére par les atoms, *Atti Congr. Intern. Fisici Como*, vol. 2, pp. 545–557.
66. E. H. Kronheimer and R. Penrose (1967). On the Structure of Causal Spaces, *Proc. Camb. Phil. Soc.*, **63**, pp. 481–501.
67. R. de L. Kronig (1926). On the theory of dispersion of X-rays, *J. Opt. Soc. Am. and Rev. Sci. Instr.*, **12**, pp. 547–557.
68. S. Lang (1972). *Differential Manifolds*, Addison-Wesley, Reading, MA.
69. H. B. Lawson Jr. (1985). *The Theory of Gauge Fields in Four Dimensions*, Conference Board of the Mathematical Sciences, Regional Conference Series in Mathematics, No. 58, published by the American Mathematical Society, Providence, RI.
70. T. Levi-Civita (1917). Nozione de parallelismo in una varietà qualunque e conseguente specificazione geometrica della curvatura Riemanniana, *Rendiconti di Palermo*, **42**, pp. 173–205.
71. L. H. Loomis (1953). *Abstract Harmonic Analysis*, Van Nostrand, Princeton, NJ.
72. G. E. Martin (1975). *The Foundations of Geometry and the Non-Euclidean Plane*, Undergraduate Texts in Mathematics, Springer-Verlag, New York.
73. W. S. Massey (1977). *Algebraic Topology: An Introduction*, Springer-Verlag, Graduate Texts in Mathematics vol. 56, New York.
74. W. S. Massey (1980). *Singular Homology Theory*, Springer-Verlag, Graduate Texts in Mathematics vol. 70, New York.
75. Y. Matsushima (1973). *Differential Manifolds*, Marcel Dekker, New York.
76. J. C. Maxwell (1954). *A Treatise on Electricity and Magnetism*, 3rd ed., Reprinted by Dover Publications, New York.
77. J. R. Munkres (1975). *Topology: A First Course*, Prentice-Hall, Englewood Cliffs, NJ.
78. J. R. Munkres (1984). *Elements of Algebraic Topology*, Addison-Wesley, Menlo Park, CA.
79. L. O'Raifeartaigh, Ed. (1972). *General Relativity: Papers in honour of J. L. Synge*, Oxford University Press, Oxford.
80. W. Pauli (1958). *Theory of Relativity*, translated from the German by G. Feld, with Supplementary Notes by the author, Pergamon Press, London. Originally published as the article "Relativitätstheorie", in *Encyklopaedie der mathematischen Wissenschaften*, vol. V19, Teubner, Leipzig, 1921.
81. G. K. Pedersen (1979). *C*-Algebras and their Automorphism Groups*, Academic Press, London.
82. R. Penrose (1972). *Techniques of Differential Topology in Relativity*, Regional Conference Series in Applied Mathematics, SIAM, Philadelphia, PA.
83. A. A. Penzias and R. W. Wilson (1965). A Measurement of Excess Antenna Temperature at 4080 MC/s, *Astrophys. J.*, **142**, pp. 419–421.
84. K. R. Popper (1959). *The Logic of Scientific Discovery*, Hutchinson Education. Reprinted by Routledge, London, 1995.
85. K. R. Popper (1957). *The Open Society and its Enemies* (in two volumes), 3rd revised edition, Routledge and Kegan Paul, London.

86. M. Reed and B. Simon (1972). *Methods of Modern Mathematical Physics*, I. Functional Analysis, Academic Press, New York.

87. K.-H. Rehren (2000). Local Quantum Observables in the Anti-de Sitter—Conformal QFT Correspondence, *Phys. Lett.*, **B 493**, pp. 383–388.

88. K.-H. Rehren (2000). Algebraic Holography, *Ann. Henri Poincaré*, **1**, pp. 607–623.

89. F. Riesz and B. Sz.-Nagy (1955). *Functional Analysis*, Frederick Ungar, New York.

90. W. Rindler (2001). *Relativity: Special, General and Cosmological*, Oxford University Press, Oxford.

91. W. Rudin (1973), *Functional Analysis*, McGraw-Hill, New York.

92. H. H. Schaefer (1971). *Topological Vector Spaces*, Springer-Verlag, New York.

93. E. Scholz, Ed. (2001). *Hermann Weyl's* Raum-Zeit-Materie *and a General Introduction to his Scientific Work*, DMV Seminar, Band 30, Birkhäuser, Basel.

94. J. Schröter (1988). An Axiomatic Basis of Space-Time Theory, I: Construction of a causal space with coordinates, *Rep. Math. Phys.*, **26**, pp. 303–333.

95. J. W. Schutz (1997). *Independent Axioms for Minkowski Space-time*, Pitman Research Notes in Mathematics, vol. 373, Addison Wesley Longman, Harlow, England.

96. R. N. Sen (1999). Why is the Euclidean Line the same as the Real Line?, *Foundations of Physics Letters*, **12**, p. 325.

97. R. N. Sen and A. Gersten, Eds. (1994). *Mathematical Physics Towards the 21st Century*, Ben-Gurion University of the Negev Press, Beer Sheva, Israel.

98. T. Shirota (1952). A Class of Topological Spaces, *Osaka Math. J.*, **4**, pp. 23–40.

99. G. F. Simmons (1963). *Introduction to Topology and Modern Analysis*, McGraw-Hill, New York.

100. E. H. Spanier (1966). *Algebraic Topology*, McGraw-Hill, New York.

101. L. A. Steen and J. A. Seebach, Jr. (1978). *Counterexamples in Topology*, 2nd ed., Springer-Verlag, New York.

102. N. Steenrod (1956). *The Topology of Fiber Bundles*, Princeton University Press, Princeton.

103. M. Steiner (1998). *The Applicability of Mathematics as a Philosophical Problem*, Harvard University Press, Cambridge, MA.

104. S. Sternberg (1964). *Lectures on Differential Geometry*, Prentice-Hall, Englewood Cliffs, NJ.

105. J. W. Strutt, Baron Rayleigh (1945). *The Theory of Sound*: One-volume reprint by Dover Publications, New York. Originally published in two volumes, vol. 1, 1877, vol. 2, 1888, London.

106. J. L. Synge (1965). *Relativity, The Special Theory*, 2nd edition, North-Holland, Amsterdam.

107. J. L. Synge (1971). *Relativity, The General Theory*, North-Holland, Amsterdam.

108. J. W. Tukey (1940). *Convergence and Uniformity in Topology*, Ann. of Math. Studies **2**, Princeton University Press, Princeton, NJ.

109. A. Weil (1937). *Sur les espaces a structure uniforme et sur la topologie générale*, Actualites Sci. et Ind. **551**, Hermann, Paris.

110. S. Weinberg (1972). *Gravitation and Cosmology*: Principles and Applications of the General Theory of Relativity, John Wiley, New York.

111. J. Wess (2003). Deformed Coordinate Spaces: Derivatives, Lecture given at the Perspectives of Balkans Collaboration, Vrnjacka Banja, Serbia, 29 Aug–2 Sep, **hep-th/0408080**.

112. H. Weyl (1913). *Die Idee der Riemannschen Fläche*, B. G. Teubner, Leipzig and Berlin; 3rd edition, 1955. English translation (of the 3rd edition) by G. R. MacLane, *The Concept of a Riemann Surface*, Addison-Wesley, Reading, MA, 1955.

113. H. Weyl (1918). Reine Infinitesimalgeometrie, Math. Zeits. **2**, pp. 384–411 (reprinted in [121], vol. II).

114. H. Weyl (1921). Zur Infinitesimalgeometrie: Einordnung der projektiven und konformen Auffassung, *Nachrichten der Königlichen Gesellschaft der Wissenschaften zu Göttingen, Mathematische-physikalische Klasse*, 99–112. Reprinted in [121], vol. II, and in [122].

115. H. Weyl (1922). *Space–Time–Matter*, translated by Henry L. Brose from the 4th German Edition (Berlin, 1922), Methuen, London. Reprinted by Dover Publications, New York.

116. H. Weyl (1923). *Raum–Zeit–Materie*, 5th Edition, Julius Springer, Berlin; a 7th edition, with additions and corrections by J. Ehlers, was published by Springer-Verlag, Berlin, in 1988.

117. H. Weyl (1923). *Mathematische Analyse des Raumproblems*, Julius Springer, Berlin.

118. H. Weyl (1926). Philosophie der Mathematik und Naturwissenschaft, in: *Handbuch der Philosophie*, R. Oldenbourg, München.

119. H. Weyl (1929). On the Foundations of General Infinitesimal Geometry, *Bull. Amer. Math. Soc.*, **35**, 716–725 (reprinted in [121], vol. III).

120. H. Weyl (1949). *Philosophy of Mathematics and Natural Science*, Princeton University Press, Princeton, NJ, U.S.A. (Revised and augmented English edition of [118].)

121. H. Weyl (1968). *Gesammelte Abhandlungen* (Collected Papers), Bd. I-IV, Springer-Verlag, Berlin.

122. H. Weyl (1956). *Selecta Hermann Weyl*, Birkhäuser Verlag, Basel.

123. H. Whitney (1936). Differentiable Manifolds, *Annals of Math.*, **37**, 645–680.

124. G. T. Whyburn (1942). *Analytic Topology*, American Mathematical Society (Colloquium Publications, vol. 28), Providence, RI.

125. G. T. Whyburn (1958). *Topological Analysis*, Princeton University Press, Princeton.

126. A. S. Wightman and R. F. Streater (1968). *PCT, Spin and Statistics, and All That*, Benjamin, New York.

127. A. S. Wightman (1976). Hilbert's Sixth Problem: Mathematical Treatment of the Axioms of Physics, in [15].

128. E. P. Wigner (1960). The Unreasonable Effectiveness of Mathematics in the Natural Sciences, *Comm. Pure and Appl. Math.*, **13**, 1–14. Reprinted in [129].

129. E. P. Wigner (1967). *Symmetries and Reflections: Scientific Essays*. Edited by W. J. Moore and M. Scriven, Indiana University Press, Bloomington, IN. Reprinted by the M.I.T. Press, Cambridge, MA, 1970. Reference [128] appears on pp. 222–237 of this volume.

130. S. Willard (1970). *General Topology*, Addison-Wesley, Reading, MA, USA.

131. N. M. J. Woodhouse (1973). The Differentiable and Causal Structures of Space-Time, *J. Math. Phys.*, **14**, 495–501.

132. E. C. Zeeman (1964). Causality Implies the Lorentz Group, *J. Math. Phys.*, **5**, pp. 490–493.

133. E. C. Zeeman (1967). The Topology of Minkowski Space, *Topology*, **6**, pp. 161–170.

List of Symbols

Symbol	Page	Description
$^l{>}\,,\, {<}^l$		Total order on light rays
on l	16	
on $\breve{l}$	74	
on ℓ	88	
${<}^{ll}$		Nonreflexive, nonsymmetric order on light rays
on l	16	
on $\breve{l}$	74	
on ℓ	88	
${>}\,,\,{<}$		Causal order on the entire space
on M	24	
on $\breve{M}$	76	
$\ll$		Nonreflexive timelike order
on M	29	
on $\breve{M}$	77	
$\gg$		Nonreflexive timelike order
on M	30	
on $\breve{M}$	77	
$\breve{\partial}$	75	Mantle operator
$\square$	17	End of definition
$\blacksquare$	18	End of proof
C_x^+	22	Forward (future) cone at x
C_x^-	22	Backward (past) cone at x
C_x	23	Cone at x
$\beta C_x^{\pm}$	27	β-boundaries of the cones at x
$\tau C_x^{\pm}$	27	τ-interiors of the cones at x
$C_{a;D}^{\pm}$	74	Local cones at $a \in M$
$\beta C_{a;D}^{\pm}$	74	β-boundaries of local cones at a

Index

Lecture Notes in Physics

For information about earlier volumes
please contact your bookseller or Springer
LNP Online archive: springerlink.com

Printing: Krips bv, Meppel
Binding: Stürtz, Würzburg